Lecture Notes in Computer Science   8600

Commenced Publication in 1973
Founding and Former Series Editors:
Gerhard Goos, Juris Hartmanis, and Jan van Leeuwen

T0234541

Christian Blum   Gabriela Ochoa (Eds.)

Evolutionary Computation in Combinatorial Optimization

14th European Conference, EvoCOP 2014
Granada, Spain, April 23-25, 2014
Revised Selected Papers

 Springer

Volume Editors

Christian Blum
IKERBASQUE, Basque Foundation for Science
University of the Basque Country
Department of Computer Science and Artificial Intelligence
Paseo Manuel Lardizabal 1, 20018 San Sebastian, Spain
E-mail: christian.c.blum@gmail.com

Gabriela Ochoa
University of Stirling, School of Natural Sciences
Department of Computing Science and Mathematics
Cottrell Building, Stirling FK9 4LA, UK
E-mail: gabriela.ochoa@cs.stir.ac.uk

Cover illustration designed by Laura Pirovano.

ISSN 0302-9743   e-ISSN 1611-3349
ISBN 978-3-662-44319-4   e-ISBN 978-3-662-44320-0
DOI 10.1007/978-3-662-44320-0
Springer Heidelberg New York Dordrecht London

Library of Congress Control Number: 2014944311

LNCS Sublibrary: SL 1 – Theoretical Computer Science and General Issues

Typesetting: Camera-ready by author, data conversion by Scientific Publishing Services, Chennai, India

Printed on acid-free paper

Springer is part of Springer Science+Business Media (www.springer.com)

Preface

During past decades, metaheuristic algorithms have been shown to be provenly effective for a wide range of hard combinatorial optimization problems arising in a variety of industrial, economic, and scientific settings. Well-known examples of metaheuristics include, but are not limited to, ant colony optimization, evolutionary algorithms, greedy randomized adaptive search procedures, iterated local search, simulated annealing, tabu search and variable neighborhood search. Metaheuristics have been applied to many different types of optimization problems, including scheduling, timetabling, network design, transportation and distribution, vehicle routing, packing and cutting, satisfiability and general integer linear programing. The series of EvoCOP events is dedicated, in particular, to algorithmic advances in this field of research.

The first edition of EvoCOP was held in 2001. Since then the event has been held annually. Notably, EvoCOP was the first event specifically dedicated to the application of evolutionary computation and related methods to combinatorial optimization problems. Originally held as a workshop, EvoCOP eventually became a conference in 2004. Past events gave researchers an excellent opportunity to present their latest research and to discuss current developments and applications. Following the general trend of the disappearance of boundaries between different metaheuristics, EvoCOP has broadened its scope in recent years and has solicited papers on any kind of metaheuristic for combinatorial optimization.

This volume contains the proceedings of EvoCOP 2014, the 14th European Conference on Evolutionary Computation in Combinatorial Optimization. It was held in Granada, Spain, during April 23–25, 2014, jointly with EuroGP 2014, the 17th European Conference on Genetic Programming, EvoBIO 2014, the 12th European Conference on Evolutionary Computation, Machine Learning and Data Mining in Computational Biology, EvoMUSART 2014, the Third International Conference on Evolutionary and Biologically Inspired Music, Sound, Art and Design, and EvoApplications 2014 (formerly EvoWorkshops), which consisted of 13 individual tracks ranging from complex systems over evolutionary algorithms in energy applications to evolutionary robotics. Since 2007, all these events are grouped under the collective name EvoStar, and constitute Europe's premier co-located event on evolutionary computation and metaheuristics.

Accepted papers of previous EvoCOP editions were published by Springer in the series Lecture Notes in Computer Science (LNCS – Volumes 2037, 2279, 2611, 3004, 3448, 3906, 4446, 4972, 5482, 6022, 6622, 7245, 7832). Below we report statistics for each conference.

EvoCOP	submitted	accepted	acceptance ratio
2001	31	23	74.2%
2002	32	18	56.3%
2003	39	19	48.7%
2004	86	23	26.7%
2005	66	24	36.4%
2006	77	24	31.2%
2007	81	21	25.9%
2008	69	24	34.8%
2009	53	21	39.6%
2010	69	24	34.8%
2011	42	22	52.4%
2012	48	22	45.8%
2013	50	23	46.0%
2014	42	20	47.6%

The rigorous, double-blind reviewing process of EvoCOP 2014 resulted in the selection of 20 out of 42 submitted papers; the acceptance rate was 47.6%. Even though slightly lower, the number of submissions was in line with previous years, which is–given the current times of crisis and limited funding–a rather remarkable achievement. At this point we would like to emphasize the work of the Program Committee. In fact, the dedicated work of our Program Committee members is essential for the continuing success of EvoCOP. We would also like to mention that acceptance/rejection decisions were not only based on the received referee reports but also on a personal evaluation of the program chairs.

There are various persons and institutions that contributed to the success of the conference and to whom we would like to express our appreciation. First of all, we thank the local organizers of EvoStar 2014, J.J. Merelo and his team, from the University of Granada. They did an extraordinary job. Furthermore, we would like to thank Marc Schoenauer from Inria (France) for his continuing support concerning the MyReview conference management system. We also thank Kevin Sim from Edinburgh Napier University, Mauro Castelli from the Universidade Nova de Lisboa and Pablo García Sánchez from the University of Granada for an excellent web site and publicity material. Thanks are also due to Jennifer Willies and the Institute for Informatics and Digital Innovation at Napier University in Edinburgh, Scotland, for administrative support and event coordination. Finally, we gratefully acknowledge the University of Granada for its support to EvoStar.

Last, but not least, we would like to thank Carlos Cotta, Peter Cowling, Jens Gottlieb, Jin-Kao Hao, Jano van Hemert, Peter Merz, Martin Middendorf, and Günther R. Raidl for their hard work and dedication at past editions of EvoCOP, which contributed to making this conference one of the reference events in evolutionary computation and metaheuristics.

May 2014

Christian Blum
Gabriela Ochoa

Organization

EvoCOP 2014 was organized jointly with EuroGP 2014, EvoBIO 2014, EvoMUSART 2014, and EvoApplications 2014.

Organizing Committee

PC Chairs

Christian Blum	IKERBASQUE
	University of the Basque Country, Spain
Gabriela Ochoa	University of Stirling, UK

Local Organization

Juan J. Merelo	University of Granada, Spain
	The whole organizer team of the University of Granada

Publicity Chairs

Kevin Sim	University of Edinburgh, UK
Mauro Castelli	Universidade Nova de Lisboa, Portugal
Pablo García Sánchez	University of Granada, Spain

EvoCOP Steering Committee

Carlos Cotta	Universidad de Málaga, Spain
Peter Cowling	University of York, UK
Jens Gottlieb	SAP AG, Germany
Jin-Kao Hao	University of Angers, France
Jano van Hemert	University of Edinburgh, UK
Peter Merz	Hannover University of Applied Sciences and Arts, Germany
Martin Middendorf	University of Leipzig, Germany
Günther Raidl	Vienna University of Technology, Austria

Program Committee

Adnan Acan	Eastern Mediterranean University, Turkey
Hernán Aguirre	Shinshu University, Japan
Enrique Alba	Universidad de Málaga, Spain

Table of Contents

A Hybrid Ant Colony Optimization Algorithm for the Far From Most String Problem

Christian Blum[1,2] and Paola Festa[3]

[1] Department of Computer Science and Artificial Intelligence,
University of the Basque Country UPV/EHU, San Sebastian, Spain
`christian.blum@ehu.es`
[2] IKERBASQUE, Basque Foundation for Science, Bilbao, Spain
[3] Department of Mathematics and Applications "R. Caccioppoli",
University of Napoli FEDERICO II, Italy
`paola.festa@unina.it`

Abstract. The *far from most string problem* belongs to the family of string selection and comparison problems known as *sequence consensus problems*, where a finite set of sequences is given and one is interested in finding their consensus, that is, a new sequence that represents as much as possible all the given sequences. Among the consensus problems, the far from most string problem is computationally one of the hardest ones with applications in several fields, including molecular biology where one is interested in creating diagnostic probes for bacterial infections or in discovering potential drug targets.

This paper comes with several contributions. On one side, the first linear integer programming formulation for the considered problem is introduced. On the other side, a hybrid ant colony optimization approach for finding good approximate solution to the problem is proposed. Both approaches are compared to the current state of the art, which is a recently proposed hybrid GRASP with path-relinking. Computational results on a large set of randomly generated test instances indicate that the hybrid ACO is very competitive.

1 Introduction

The combinatorial optimization problem tackled in this paper is known as the *far from most string problem* (FFMSP). It belongs to a family of string problems labelled as *sequence consensus problems*, where a finite set of sequences is given and one is interested in finding their consensus, that is, a new sequence that represents as much as possible all the given sequences. Several different (and even opposing) objectives may be considered in the context of sequence consensus problems. Examples include the following ones.

1. The consensus is a new sequence whose total distance from all given sequences is minimal (*closest string problem*) (CSP);
2. The consensus is a new sequence whose total distance from all given sequences is maximal (*farthest string problem*) (FSP);

C. Blum and G. Ochoa (Eds.): EvoCOP 2014, LNCS 8600, pp. 1–12, 2014.

3. The consensus is a new sequence far from most of the given sequences, which is the case of the FFMSP.

Apart from introducing a linear integer programming model, the main contribution of the paper is a hybrid ant colony optimization algorithm (ACO) [2] for tackling the FFMSP. ACO algorithms are metaheuristics inspired by the shortest path finding behaviour of natural ant colonies. The hybrid algorithm outlined in this work concerns a sequential combination of ant colony optimization with a mathematical programming solver. In the following, we first provide the necessary notation for being able to state the FFMSP in a technical way. After providing the problem definition we give a short summary on related work.

1.1 Notation

In order to be able to formally state the FFMSP, the following notation is introduced.

- An alphabet $\Sigma = \{c_1, \ldots, c_k\}$ is a finite set of k different characters.
- $s^i = (s^i_1, s^i_2, \ldots, s^i_m)$ denotes a sequence of m characters (that is, of length m) over alphabet Σ.
- Given two sequences s^i and s^l over Σ such that $|s^i| = |s^l|$, the (generalized) Hamming distance $d_H(s^i, s^l)$ between s^i and s^l is calculated as follows:

$$d_H(s^i, s^l) := \sum_{j=1}^{|s^i|} \phi(s^i_j, s^l_j) \ , \tag{1}$$

  where s^i_j and s^l_j denote the character at position j in sequence s^i and in sequence s^l, respectively, and $\phi : \Sigma \times \Sigma \mapsto \{0, 1\}$ is a predicate function such that $\phi(a, b) = 0$, if $a = b$, and $\phi(a, b) = 1$, otherwise.
- Given a set $\Omega = \{s^1, \ldots, s^n\}$ of n sequences of length m over Σ, d^Ω_H denotes the minimum Hamming distance between all sequences in Ω. Formally,

$$d^\Omega_H := \min \left\{ d_H(s^i, s^l) \mid i, l \in \{1, \ldots, n\}, i < l \right\} \ . \tag{2}$$

  Note that $0 \leq d^\Omega_H \leq m$.
- Each sequence s of length m over Σ is a valid solution to the FFMSP. Given any valid solution s and a threshold value $0 \leq t \leq m$, set $P^s \subseteq \Omega$ is defined as follows:

$$P^s := \{s^i \in \Omega \mid d_H(s^i, s) \geq t\}. \tag{3}$$

1.2 Problem Definition

Given a fixed threshold value t (where $0 \leq t \leq m$), a finite alphabet Σ of size k, and a set $\Omega = \{s^1, \ldots, s^n\}$ of n sequences of length m over Σ, the goal of the FFMSP consists in finding a sequence s^* of length m over Σ such that P^{s^*}

is of maximal size. In other words, given a solution s, the objective function $f : \Sigma^m \mapsto \{1, \ldots, n\}$ is defined as follows:

$$f(s) := |P^s| \qquad (4)$$

Among the sequence consensus problems, the FFMSP is one of the computationally hardest. In fact, compared to the other consensus problems, it is much harder to approximate, due to the approximation preserving reduction to the FFMSP from the independent set problem, which is a classical and computationally intractable combinatorial optimization problem. In 2003, Lanctot et al. [11] proved that for sequences over an alphabet Σ with $|\Sigma| \geq 3$, approximating the FFMSP within a polynomial factor is NP-hard.

1.3 Related Work

As indicated by the available theoretical results on computational hardness, polynomial time algorithms for the FFMSP can yield only solutions with no constant guarantee of approximation. Moreover, with growing instance size, complete techniques quickly reach their limits, and (meta-)heuristics become the only feasible approach for deriving high-quality solutions. The first attempt in the design of efficient metaheuristic approaches has been done in 2005 by Meneses et al. [12], who proposed a heuristic algorithm consisting of a simple greedy construction followed by an iterative improvement phase. Later, in 2007 Festa [5] designed a GRASP and more recently in 2012 Festa and Pardalos [6] proposed a genetic algorithm. In 2012, Mousavi et al. [13] devised a new function to be used as an alternative to the objective function when evaluating neighbor solutions during the local search phase of the algorithm proposed in [5]. In 2013, Ferone et al. [4] developed several pure and hybrid multistart iterative heuristics, including

◇ a pure GRASP approach, inspired by [5];
◇ a GRASP making use of path-relinking for intensification;
◇ a pure variable neighborhood search (VNS) approach;
◇ a VNS using path-relinking for intensification;
◇ a GRASP that uses VNS to implement the local search phase;
◇ and a GRASP approach that uses VNS to implement the local search phase and path-relinking for intensification.

Among all the presented algorithms, the GRASP approach making use of path-relinking for intensification can currently regarded to be state of the art for obtaining very good solutions in a resonable amount of computation time. GRASP [3, 7–9] is a multistart heuristic, where at each iteration a solution is probabilistically generated biased by a greedy function. Subsequently, a local search phase is applied starting from the just built solution until a locally optimal solution is found. Repeated applications of the construction procedure yield diverse starting solutions for the local search and the best overall local optimal solution is returned as the result.

1.4   Organization of the Paper

The remainder of this article is organized as follows. In Section 2, we present the first linear integer programming formulation for the FFMSP. In Section 3, we describe a hybrid ant colony approach based on a combination of ant colony optimization and a mathematical programming solver. Computational results for both the linear integer programming model and the ACO approach are reported in Section 4. Finally, concluding remarks and an outlook to future work are given in Section 5.

2   A Linear Integer Programming Model

First, we developed a linear integer programming model for solving the FFMSP. This model works on two sets of binary variables. The first set contains for each position i $(i = 1, \ldots, m)$ of a possible solution and for each character $c_j \in \Sigma$ $(j = 1, \ldots, k)$ a binary variable $x_{i,c_j} \in \{0,1\}$. The second set consists of a binary variable $y_r \in \{0,1\}$ $(r = 1, \ldots, n)$ for each of the n input sequences provided in set Ω. The linear integer program itself can then be stated as follows.

$$\max \sum_{r=1}^{n} y_r \tag{5}$$

subject to:

$$\sum_{c_j \in \Sigma} x_{i,c_j} = 1 \quad \text{for } i = 1, \ldots, m \tag{6}$$

$$\sum_{i=1}^{m} x_{i,s_i^r} \leq m - t y_r \quad \text{for } r = 1, \ldots, n \tag{7}$$

$$x_{i,c_j}, y_r \in \{0,1\}$$

Hereby, constraints (6) ensure that for each position i of a possible solution exactly one character from Σ is chosen. Moreover, constraints (7) ensure that y_r can only be set to 1 if and only if the number of differences between $s^r \in \Omega$ and the possible solution (as defined by the setting of the variables x_{i,c_j}) is at least t. Remember, in this context, that s_i^r denotes the character at position i in $s^r \in \Omega$.

3   The Proposed Approach

This section is dedicated to a description of the hybrid ACO approach that we developed for tackling the FFMSP. In the following we first deal with preliminaries, before the framework of the proposed algorithm is outlined.

3.1 Preliminaries

Just like any other learning-based algorithm, ACO generally faces difficulties in the case of an objective function characterized by large plateaus. Unfortunately, this is exactly the case of the FFMSP, because the range of possible objective function values—that is, $\{0, \ldots, n\}$—is rather small. Therefore, we decided to introduce a refined way for comparing between valid solutions as follows. First, given a valid solution s, set $Q^s \subseteq \Omega$ is indirectly defined as follows:

- $|Q^s| = \max\{|P^s| + 1, n\}$;
- $d_H(s', s) \geq d_H(s'', s)$ for all $s' \in Q^s$ and $s'' \in \Omega \setminus Q^s$.

In words, set Q^s consists of all sequences $s' \in \Omega$ that have at least Hamming distance t with respect to s. Moreover, in case $|P^s| < n$, set Q^s also contains the sequence $\hat{s} \in \Omega \setminus P^s$ that has, among all sequence in $\Omega \setminus P^s$, the largest Hamming distance with respect to s.

Given two valid solutions s and s', a comparison operator (denoted by $>_{lex}$) is used for comparing them. This operator is defined in the following way. It holds that $s >_{\text{lex}} s'$ if and only if

$$ f(s) > f(s') \textbf{ or } \left(f(s) = f(s') \textbf{ and } \sum_{s'' \in Q^s} d_H(s'', s) > \sum_{s'' \in Q^{s'}} d_H(s'', s') \right) . \quad (8) $$

In other words, the comparison between s and s' is done in a lexicographical way with the original objective function value as a first criterion, and the sum of the Hamming distances of the solutions from the respective sets Q^s and $Q^{s'}$ with respect to s and s' as a second criterion. It is intuitively assumed that the solution for which this sum is greater is somehow closer to a solution with a higher objective function value.

The second aspect of this section concerns the definition of the pheromone model and the greedy information, both being of crucial importance for any ACO algorithm. The pheromone model $\mathcal{T}$ consists of a pheromone value τ_{i,c_j} for each combination of a position i ($i = 1, \ldots, m$) in a solution sequence, and a letter $c_j \in \Sigma$. Moreover, to each of these combinations is assigned a greedy value η_{i,c_j} which measures the desirability to assign letter c_j to position i. Intuitively, value η_{i,c_j} should be reverse-proportional to the number of occurences of letter c_j at position i in the n input sequences. That is,

$$ \eta_{i,c_j} := \frac{(n - |\{s \in \Omega \mid s_i = c_j\}|)}{n} . \quad (9) $$

3.2 Algorithmic Framework

The general framework of the proposed algorithm (see Algorithm 1) is as follows. First of all, calls to function RunACO(s^{bs}) are executed until either the time limit $tlim_{\text{ACO}}$ is reached, or the number of unsuccessful calls to RunACO(s^{bs}),

---

Algorithm 1. (Hybrid) ACO for the FFMSP

1. **input:** Ω, Σ, t, $tlim_{\text{ACO}}$, $runlim_{\text{ACO}}$, $tlim_{\text{CPLEX}}$
2. $s^{bs} :=$ NULL
3. $runs_{\text{ACO}} := 0$
4. **while** time limit $tlim_{\text{ACO}}$ not reached **and** $runs_{\text{ACO}} < runlim_{\text{ACO}}$ **do**
5. $s := \text{RunACO}(s^{bs})$
6.   **if** $s >_{\text{lex}} s^{bs}$ **then**
7. $s^{bs} := s$
8. $runs_{\text{ACO}} := 0$
9.   **else**
10. $runs_{\text{ACO}} := runs_{\text{ACO}} + 1$
11.   **end if**
12. **end while**
13. **if** time limit $tlim_{\text{CPLEX}}$ not reached **then**
14. $s^{bs} := \text{SolutionPolishing}(s^{bs}, tlim_{\text{CPLEX}})$
15. **end if**
16. **output:** the best-so-far solution s^{bs}

---

as counted by $runs_{\text{ACO}}$, has reached the limit $runlim_{\text{ACO}}$. Hereby, a call to RunACO(s^{bs}) is regarded as *unsuccessful* if the best-so-far solution s^{bs} was not improved during the run. Function RunACO(s^{bs}) takes the best solution found so far (s^{bs}) as input and applies an ACO algorithm (as outlined below) until the convergence of the pheromone values is reached.

After this first phase (lines 4–12 of Algorithm 1) the algorithm possibly applies a second phase in which the hybridization with a mathematical programming solver takes place. This hybrid algorithm component works as follows. The best solution obtained by the ACO-phase is given to the MIP-solver CPLEX as a starting solution (see, for example, page 531 of the IBM ILOG CPLEX V12.1 user manual: *Starting from a Solution: MIP Starts*). CPLEX is then used to apply *solution polishing* to the given starting solution until the computation time limit $tlim_{\text{CPLEX}}$ is reached (see, for example, page 521 of the IBM ILOG CPLEX V12.1 user manual: *Solution polishing*). This is done in function SolutionPolishing($s^{bs}, tlim_{\text{CPLEX}}$). Solution polishing can be seen as a black box local search based on branch & cut, with the aim to improve a solution, rather than proving optimality. The best solution found after this phase is provided as output of the algorithm.

Note that a hybrid algorithm is obtained by a setting of $tlim_{\text{ACO}} < tlim_{\text{CPLEX}}$. On the other side, the pure ACO approach can be applied by setting, for example, $tlim_{\text{CPLEX}} = 0$ and by assigning a very large value to $runlim_{\text{ACO}}$ (in order to avoid that the while-loop of line 4 is stopped before the computation time limit $tlim_{\text{ACO}}$ is reached).

3.3 The ACO Phase

In the following, we focus on the description of function RunACO(s^{bs}) of Algorithm 1, which is pseudo-coded in Algorithm 2. Note that the ACO-implementation

---

Algorithm 2. Function RunACO(s^{bs}) of Algorithm 1

1. **input:** s^{bs}
2. $s^{rb} :=$ NULL, $cf := 0$
3. Initialize all pheromone values of $\mathcal{T}$ to 0.5
4. **while** $cf < 0.99$ **do**
5. $s^{ib} :=$ NULL
6.   **for all** n_a artificial ants **do**
7. $s :=$ ConstructSolution($\mathcal{T}$)
8. $s :=$ LocalSearch(s)
9.   **if** $s >_{\text{lex}} s^{ib}$ **then** $s^{ib} := s$
10.   **end for**
11. $s^{ib} :=$ PathRelinking(s^{ib}, s^{bs})
12.   **if** $s^{ib} >_{\text{lex}} s^{rb}$ **then** $s^{rb} := s^{ib}$
13.   **if** $s^{ib} >_{\text{lex}} s^{bs}$ **then** $s^{bs} := s^{ib}$
14.   ApplyPheromoneUpdate($cf, \mathcal{T}, s^{ib}, s^{rb}$)
15. $cf :=$ ComputeConvergenceFactor($\mathcal{T}$)
16. **end while**
17. **output:** the best-so-far solution s^{bs}

---

used in this function is very similar to a $\mathcal{MAX}$–$\mathcal{MIN}$ Ant System ($MMAS$) implemented in the Hyper-Cube Framework (HCF) [14, 1]. The function takes as input the best-so-far solution s^{bs}. At the start, first, all pheromone values of $\mathcal{T}$ are initialized to 0.5. Then, at each iteration, $n_a = 10$ solutions are constructed in function ConstructSolution($\mathcal{T}$) on the basis of pheromone and greedy information. Local search is applied to each of these solutions in function LocalSearch(). Finally, path-relinking is applied to the best solution s^{ib} (after local search) generated in the current iteration. Solution s^{ib} is also referred to as the *iteration-best* solution. Path-relinking is applied in function PathRelinking(s^{ib}, s^{bs}). Hereby, s^{ib} serves as initial solution and the best-so-far solution s^{bs} serves as guiding solution. After updating (if necessary) the best-so-far solution and the so-called *restart-best* solution, which is the best solution found so far during the current application of RunACO(), the pheromone update is performed in function ApplyPheromoneUpdate($cf, \mathcal{T}, s^{ib}, s^{rb}$). Finally, at the end of each iteration, the new convergence factor value cf is computed in function ComputeConvergenceFactor($\mathcal{T}$). In the following we outline the working of the different functions in more detail.

ConstructSolution($\mathcal{T}$): For the construction of a new solution s, a letter from Σ is chosen successively for all positions from 1 to m. The letter choice for each position $1 \leq i \leq m$ is hereby performed as follows. First, for each letter $c_j \in \Sigma$ a choice probability $\mathbf{p}_{i,c_j}$ for position i is calculated as follows:

$$\mathbf{p}_{i,c_j} := \frac{\tau_{i,c_j} \cdot \eta_{i,c_j}}{\sum_{c \in \Sigma} \tau_{i,c} \cdot \eta_{i,c}} \qquad (10)$$

Then, a value z is chosen uniformly at random from $[0.5, 1.0]$. In case $z \leq d_{rate}$, the letter $c \in \Sigma$ with the largest probability value is chosen for position i. Otherwise, a letter $c \in \Sigma$ is chosen randomly with respect to the probability values. Hereby, d_{rate} is a parameter of the algorithm, which was set to 0.8 after tuning by hand.

LocalSearch(s) and PathRelinking(s^{ib}, s^{bs}): The local search and path-relinking procedures used in this work are the same as the ones developed in the context of the GRASP algorithm published in [4]. Therefore, we only provide a short text-based description. For the pseudo-code we refer the interested reader to the original work.

The neighborhood of a solution s considered for the local search consists in all solutions that can be obtained by ex-changing exactly one character with a different one. The positions of a solution s are examined in a certain order, and, as soon as an improving neighbor is found, this neighbor is accepted as new current solution. The procedure stops once no improving neighbor can be found. The order in which positions are examined is randomly chosen for each application of local search, which is actually the only difference of our implementation in comparison to the one as given in [4].

Path-relinking is a concept initially proposed in [10] for the search intensification between two solutions. The path-relinking procedure that we applied in function PathRelinking(s^{ib}, s^{bs}) works roughly as follows. Solution s^{ib} is used as initial solution, and the best-so-far solution s^{bs} as guiding solution. The positions that both solutions have in common remain untouched. The solution space spanned by the positions that are different in the two solutions is explored by generating a path in the solution space linking solution s^{ib} with s^{bs}. The best solution found on this path is provided as output of path-relinking.

ApplyPheromoneUpdate($cf, \mathcal{T}, s^{ib}, s^{rb}$): The ACO procedure makes use of at most two different solutions for updating the pheromone values, namely solutions s^{ib} and s^{rb}. The weight of each solution for the pheromone update is determined as a function of cf, the convergence factor. The pheromone values τ_{i,c_j} are updated as follows:

$$\tau_{i,c_j} := \tau_{i,c_j} + \rho \cdot (\xi_{i,c_j} - \tau_{i,c_j}) \ , \tag{11}$$

where

$$\xi_{i,c_j} := \kappa_{ib} \cdot \Delta(s_i^{ib}, c_j) + \kappa_{rb} \cdot \Delta(s_i^{rb}, c_j) \ . \tag{12}$$

Hereby, function $\Delta(s, c_j)$ evaluates to 1 in case character c_j is to be found at position i of solution s. Otherwise, the function evaluates to 0. Moreover, κ_{ib} is the weight of solution s^{ib} and κ_{rb} the one of s^{rb}. It is required that $\kappa_{ib} + \kappa_{rb} = 1$. The weight values that we chose are the standard ones shown in Table 1. Finally, note that the algorithm works with upper and lower bounds for the pheromone values, that is, $\tau_{max} = 0.999$ and $\tau_{min} = 0.001$. In case a pheromone values surpasses one of these limits, the value is set to the corresponding limit. This has the effect that a complete convergence of the algorithm is avoided.

Table 1. Setting of κ_{ib} and κ_{rb} depending on the convergence factor cf

	$cf < 0.4$	$cf \in [0.4, 0.6)$	$cf \in [0.6, 0.8)$	$cf \geq 0.8$
κ_{ib}	1	2/3	1/3	0
κ_{rb}	0	1/3	2/3	1

ComputeConvergenceFactor($\mathcal{T}$): The formula that was used for computing the value of the convergence factor is as follows:

$$cf := 2 \left(\left(\frac{\sum\limits_{\tau \in \mathcal{T}} \max\{\tau_{\max} - \tau, \tau - \tau_{\min}\}}{|\mathcal{T}| \cdot (\tau_{\max} - \tau_{\min})} \right) - 0.5 \right)$$

This implies that at the start of function RunACO(s^{bs}), cf has value zero. On the other side, in the case in which all pheromone values are either at $\tau_{\min}$ or at $\tau_{\max}$, cf has a value of one. In general, cf moves in $[0, 1]$. This completes the description of the proposed algorithm.

4   Experimental Evaluation

We implemented the proposed algorithm in ANSI C++ using GCC 4.7.3 for compiling the software. Moreover, the mathematical program outlined in Section 2 was solved with IBM ILOG CPLEX V12.1. The same version of CPLEX was used within the hybrid ACO approach. The experimental results that we outline in the following were obtained on a cluster of PCs with "Intel(R) Xeon(R) CPU 5160" CPUs of 4 nuclii of 3000 MHz and 4 Gigabyte of RAM.

4.1   Problem Instances

For the purpose of comparing to the state of the art, the algorithms proposed in this paper were applied to the set of benchmark instances introduced in [4]. This set consists of random instances of different size. More specifically, the number of input sequences (n) is in $\{100, 200\}$, and the length of the input sequences (m) is in $\{300, 600, 800\}$. In all cases, the alphabet size is four, that is, $|\Sigma| = 4$. For each combination of n and m, the set consists of 100 random instances. This makes a total of 600 instances. Finally, as in [4], our algorithms were applied to all instances for different settings of parameter t. In particular, $t \in \{0.75m, 0.8m, 0.85m\}$.

4.2   Results

We applied a pure version of ACO ($tlim_{\mathrm{ACO}} = 90s$, $tlim_{\mathrm{CPLEX}} = 0s$, $runlim_{\mathrm{ACO}} =$ very large integer) and a hybrid version of ACO ($tlim_{\mathrm{ACO}} = 60s$, $tlim_{\mathrm{CPLEX}} = 90s$, $runlim_{\mathrm{ACO}} = 10$) exactly once to each problem instances. This was done for each of the three considered values for t. The (total) computation time limit

Table 2. Numerical results

n, m, t	GRASP+PR	CPLEX	ACO	HyACO
100, 300, $t = 0.75m$	100	100	100	100
100, 300, $t = 0.8m$	**79.61**	69.21	73.55	77.84
100, 300, $t = 0.85m$	13.18	22.08	24.64	**28.3**
100, 600, $t = 0.75m$	100	100	100	100
100, 600, $t = 0.8m$	**80.13**	66.95	69.12	72.97
100, 600, $t = 0.85m$	4.98	19.38	20.82	**22.82**
100, 800, $t = 0.75m$	100	100	100	100
100, 800, $t = 0.8m$	**82.64**	67.28	67.43	70.94
100, 800, $t = 0.85m$	1.84	18.3	19.84	**21.66**
200, 300, $t = 0.75m$	200	200	199.38	200
200, 300, $t = 0.8m$	100	75.32	**104.3**	104.17
200, 300, $t = 0.85m$	11.9	19.16	27.1	**28.59**
200, 600, $t = 0.75m$	200	200	199.99	200
200, 600, $t = 0.8m$	**88.49**	59.29	85.53	85.02
200, 600, $t = 0.85m$	2.42	18.12	21.03	**21.9**
200, 800, $t = 0.75m$	200	200	199.99	200
200, 800, $t = 0.8m$	73.08	54.31	**78.54**	77.95
200, 800, $t = 0.85m$	0.21	18.56	19.14	**20.4**

for each run was set to 90 CPU seconds in order to be comparable to the results reported for the GRASP hybridized with path-relinking in [4]. These three algorithms are henceforth referred to by ACO, HyACO, and GRASP+PR. Moreover, the mathematical programming solver CPLEX was applied with a CPU time limit of 90 seconds to each problem instance. Note that we report the values of the best feasible solutions found by CPLEX within the available computation time. The corresponding solutions are (expect for the case $t = 0.75m$) not proven to be optimal ones. The results are reported in Table 2. Note that for each combination of n, m, and t the given values are averages over the results for 100 problem instances. In each case, the best results are shown in bold font. For the analysis of the results, the three cases resulting from $t \in \{0.75m, 0.8m, 0.85m\}$ are treated seperately. This is because the behaviour of the algorithms changes greatly from one case to another.

Case $t = 0.75m$: This case results clearly in being the easiest one for all algorithms. In fact, GRASP+PR, CPLEX and HyACO find optimal solution to all 600 problem instances. The only algorithm that has slight difficulties when $n = 200$ is the pure ACO approach, which does not find an optimal solution in a few cases.

Case $t = 0.8m$: In this case, CPLEX is not able to prove optimality for any of the 600 instances. In fact, especially for the larger instances—that is, instances with $n = 200$—the performance of CPLEX decreases strongly. Concerning a comparison of the two ACO version the following can be observed. HyACO appears to outperform ACO consistently for the smaller instances ($n = 100$). This is not the case anymore for the larger instances, where the performance of the

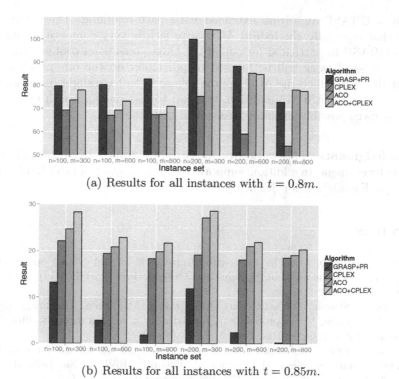

(a) Results for all instances with $t = 0.8m$.

(b) Results for all instances with $t = 0.85m$.

Fig. 1. Graphical representation of the results from Table 2 for $t = 0.8m$ (see (a)) and $t = 0.85m$ (see (b))

two algorithms is comparable. GRASP+PR outperforms both ACO approaches on the set of smaller instances. However, this seems to change with growing instance size. In fact, when $n = 200$ and $m \in \{300, 800\}$ both ACO approaches outperform GRASP+PR. The results for this case are also shown in graphical form in Figure 1(a).

Case $t = 0.85m$**:** The results for this case are quite surprising. In fact, comparing the results of GRASP+PR with the results of the three new approaches, it turns out that GRASP+PR is not at all able to solve this case (see also the graphical representation of the results in Figure 1(b). Concerning a comparison of the three new approaches, it appears that HyACO consistently outperforms the pure ACO approach, which, in turn, consistently outperforms CPLEX.

5   Conclusions and Future Work

In this work we introduced a pure and a hybrid ACO approach for tackling the so-called *far from most string* problem. Moreover, we developed the first linear integer programming model for this problem. Finally, we described an experimental evaluation and a comparison to the state of the art from the literature, which is a GRASP algorithm extended with path-relinking. The comparison showed that especially the hybrid ACO approach is very competitive, outperforming GRASP in particular for one of the three considered problem cases.

Future lines of work will be focused in particular on the one problem case (out out three) in which the ACO approaches seem to have slight disadvantages with respect to GRASP. In addition, we plan to test out approach on a larger set of instances possibly including real-world instances.

Acknowledgments. This work was supported by grant TIN2012-37930 of the Spanish Government. In addition, support is acknowledged from IKERBASQUE, the Basque Foundation for Science.

References

1. Blum, C., Dorigo, M.: The hyper-cube framework for ant colony optimization. IEEE Transactions on Man, Systems and Cybernetics – Part B 34(2), 1161–1172 (2004)
2. Dorigo, M., Stützle, T.: Ant Colony Optimization. MIT Press, Cambridge (2004)
3. Feo, T., Resende, M.: A probabilistic heuristic for a computationally difficult set covering problem. Oper. Res. Lett. 8, 67–71 (1989)
4. Ferone, D., Festa, P., Resende, M.: Hybrid metaheuristics for the far from most string problem. In: Blesa, M.J., Blum, C., Festa, P., Roli, A., Sampels, M. (eds.) HM 2013. LNCS, vol. 7919, pp. 174–188. Springer, Heidelberg (2013)
5. Festa, P.: On some optimization problems in mulecolar biology. Mathematical Bioscience 207(2), 219–234 (2007)
6. Festa, P., Pardalos, P.: Efficient solutions for the far from most string problem. Annals of Operations Research 196(1), 663–682 (2012)
7. Festa, P., Resende, M.: GRASP: An annotated bibliography. In: Ribeiro, C., Hansen, P. (eds.) Essays and Surveys on Metaheuristics, pp. 325–367. Kluwer Academic Publishers (2002)
8. Festa, P., Resende, M.: An annotated bibliography of GRASP – Part I: Algorithms. International Transactions in Operational Research 16(1), 1–24 (2009)
9. Festa, P., Resende, M.: An annotated bibliography of GRASP – Part II: Applications. International Transactions in Operational Research 16(2), 131–172 (2009)
10. Glover, F., Laguna, M., Martí, R.: Fundamentals of scatter search and path relinking. Control and Cybernetics 39, 653–684 (2000)
11. Lanctot, J., Li, M., Ma, B., Wang, S., Zhang, L.: Distinguishing string selection problems. Information and Computation 185(1), 41–55 (2003)
12. Meneses, C., Oliveira, C., Pardalos, P.: Optimization techniques for string selection and comparison problems in genomics. IEEE Engineering in Medicine and Biology Magazine 24(3), 81–87 (2005)
13. Mousavi, S., Babaie, M., Montazerian, M.: An improved heuristic for the far from most strings problem. Journal of Heuristics 18, 239–262 (2012)
14. Stützle, T., Hoos, H.H.: $\mathcal{MAX}$-$\mathcal{MIN}$ Ant System. Future Generation Computer Systems 16(8), 889–914 (2000)

A Parametric Framework
for Cooperative Parallel Local Search

Danny Munera[1], Daniel Diaz[1], Salvador Abreu[2], and Philippe Codognet[3]

[1] University of Paris 1-Sorbonne, France
Danny.Munera@malix.univ-paris1.fr, Daniel.Diaz@univ-paris1.fr
[2] Universidade de Évora and CENTRIA, Portugal
spa@di.uevora.pt
[3] JFLI-CNRS / UPMC / University of Tokyo, Japan
codognet@is.s.u-tokyo.ac.jp

Abstract. In this paper we address the problem of parallelizing local search. We propose a general framework where different local search engines cooperate (through communication) in the quest for a solution. Several parameters allow the user to instantiate and customize the framework, like the degree of intensification and diversification. We implemented a prototype in the X10 programming language based on the adaptive search method. We decided to use X10 in order to benefit from its ease of use and the architectural independence from parallel resources which it offers. Initial experiments prove the approach to be successful, as it outperforms previous systems as the number of processes increases.

1 Introduction

Constraint Programming is a powerful declarative programming paradigm which has been successfully used to tackle several complex problems, among which many combinatorial optimization ones. One way of solving problems formulated as a Constraint Satisfaction Problem (CSP) is to resort to *Local Search* methods [13,12], which amounts to the methods collectively designated as *Constraint-Based Local Search* [18]. One way to improve the performance of Local Search Methods is to take advantage of the increasing availability of parallel computational resources. Parallel implementation of local search meta-heuristics have been studied since the early 90's, when multiprocessor machines started to become widely available, see [24]. One usually distinguishes between single-walk and multiple-walk methods. Single-walk methods consist in using parallelism inside a single search process, e.g., for parallelizing the exploration of the neighborhood, while multiple-walk methods (also called multi-start methods) consist in developing concurrent explorations of the search space, either independently or cooperatively with some communication between concurrent processes. A key point is that independent multiple-walk (IW) methods are the easiest to implement on parallel computers and can in theory lead to linear speed-up, cf. [24].

Previous work on independent multi-walk local search in a massively parallel context [2,7,8] achieves good but not ideal parallel speedups. On structured

C. Blum and G. Ochoa (Eds.): EvoCOP 2014, LNCS 8600, pp. 13–24, 2014.

constraint-based problems such as (large instances of) Magic Square or All-Interval, independent multiple-walk parallelization does not yield linear speedups, reaching for instance a speedup factor of "only" 50-70 for 256 cores. However on the Costas Array Problem, the speedup can be linear, even up to 8000 cores [8]. On a more theoretical level, it can be shown that the parallel behavior depends on the *sequential runtime distribution* of the problem: for problems admitting an exponential distribution, the speedup can be linear, while if the runtime distribution is shifted-exponential or (shifted) lognormal, then there is a bound on the speedup (which will be the asymptotic limit when the number of cores goes to infinity), see [23] for a detailed analysis of these phenomena.

In order to improve the independent multi-walk approach, a new paradigm that includes *cooperation* between walks has to be defined. Indeed, *Cooperative Search* methods add a communication mechanism to the IW strategy, to share or exchange information between solver instances during the search process. However, developing an efficient cooperative method is a very complex task, cf. [6], and many issues must solved: *What information is exchanged? Between what processes is it exchanged? When is the information exchanged? How is it exchanged? How is the imported data used?* [22].

We recently started to work towards a redesigned platform for parallel local search, for which some early results are described in [17]. In the present article we progress towards a general framework, while extending the experimental evaluation to a distributed computing platform.

In this article, we propose a general framework for cooperative search, which defines a flexible and parametric cooperative strategy based on the cooperative multi-walk (CW) scheme. This framework is oriented towards distributed architectures based on clusters of nodes, with the notion of "teams" running on nodes and regrouping several search engines (called "explorers") running on cores, and the idea that all teams are distributed and thus have limited inter-node communication. This framework allows the programmer to define aspects such as the degree of *intensification* and *diversification* present in the parallel search process. A good trade-off is essential to achieve good performance. For instance, a parallel scheme has been developed in [1] with groups of parallel SAT solvers communicating their best configurations on restart, but performance degrades when groups contain more than 16 processes. In [15] another approach is described where a hybrid intensification/diversification is shown to help when scaling into hundreds of cores.

We also propose an implementation of our general cooperative framework and perform an experimental performance evaluation over a set of well-known CSPs. We compare its performance against the Independent Walk implementation and show that in nearly all examples we achieve better performance. Of course, these are just preliminary results and even better performance could be obtained by optimizing the current version. An interesting aspect of the implementation it that we use the X10 programming language, a novel language for parallel processing developed by IBM Research, because it gives us more flexibility than using a more traditional approach, e.g., an MPI communication package.

The rest of the paper is organized as follow. We briefly review the adaptive local search method and present the independent Multi-Walks experiments in section 2. We introduce our Cooperative Search framework in section 3 and, subsequently, present an implementation of this framework in the X10 language in section 4. Section 5 compares the results obtained with both the Independent Multi-Walks implementation and the Cooperative Search implementation. Finally, in section 6, we conclude and propose some ideas for future work.

2   Local Search and Parallelism

In this study, we use a generic, domain-independent constraint-based local search method named Adaptive Search [3,4]. This metaheuristic takes advantage of the CSP formulation and makes it possible to structure the problem in terms of variables and constraints and to analyze the current assignment of variables more precisely than an optimization of a global cost function e.g., the number of constraints that are not satisfied. Adaptive Search also includes an adaptive memory inspired in Tabu Search [10] in which each variable leading to a local minimum is marked and cannot be chosen for the next few iterations. A local minimum is a configuration for which none of the neighbors improve the current configuration. The input of the Adaptive Search algorithm is a CSP, for each constraint an error function is defined. This function is a heuristic value to represent the degree of satisfaction of a constraint and gives an indication on how much the constraint is violated. Adaptive Search is based on iterative repair from the variables and constraint error information, trying to reduce the error in the worse variable. The basic idea is to calculate the error function for each constraint, and then combine for each variable the errors of all constraints in which it appears, thus projecting constraint errors on involved variables. Then, the algorithm chooses the variable with the maximum error as a "culprit" and selects it to modify later its value. The purpose is to select the best neighbor move for the culprit variable, this is done by considering all possible changes in the value of this variable (neighbors) and selecting the lower value of the overall cost function. Finally, the algorithm also includes partial resets in order to escape stagnation around local minimum; and it is possible to restart from scratch when the number of iterations becomes too large.

Independent Multi-Walks

To take advantage of the parallelism in Local Search methods different strategies have been proposed like functional parallelism and data parallelism. Functional parallelism aims to parallelize the search algorithm but it generally has too big overheads due to the management of the fine-grained tasks (creation, synchronization and termination) [17]. In contrast, data parallelism tries to parallelize the exploration of the search space. A straightforward implementation of data parallelism is the Independent Multi-Walks (IW) approach. The idea is to use isolated sequential Local Search solver instances dividing the search space of the problem through different random starting points [24]. This approach has been successfully used in constraint programming problems reaching good performance [2,7,8].

We implemented a IW strategy for the Adaptive Search. This implemention is developed with the PGAS language X10. We tested it on a set of 4 classical benchmarks. Three of them are taken from CSPLib [9]: the All-Interval Problem (AIP, prob007) with size 400, Langford's Numbers Problem (LNP, prob024) with size 500 and the Magic Square Problem (MSP, prob019) with size 200 × 200. The last benchmark is the Costas Array Problem [14] (CAP) with size 20. For all problems, we select difficult instances involving very large search spaces. These instances are generally out of reach of the traditional complete solvers like Gecode [21].

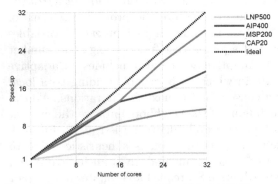

Fig. 1. Speed-ups of Independent Multi-Walks on a distributed system

The testing environment used in each running was a mixed cluster with 5 AMD nodes and 3 Intel nodes. Each AMD node has two quad-core Opteron 2376 processors. Each Intel node has two quad-core Xeon X3450 processors. All systems use a dedicated Gigabit-Ethernet interconnect. Figure 1 shows the speed-ups obtained when increasing the numbers of cores. We solve the instances using 8, 16, 24 and 32 cores.

The results show quasi-linear speed-ups for the CAP instance, in accordance with [8]. However, for MSP and AIP the speed-up tends to flatten out when increasing the number of cores. For instance, for MSP the speed-up is only improved by 1 unit when going from 16 to 32 cores. Finally for LNP the performance is very poor with a speed-up of 2 using 32 cores.

3   Cooperative Search Framework

As seen above, the speed-ups obtained with the IW strategy are good with few compute instances, however when the number of cores increases the performance tends to taper off and the gain is not significant. To tackle this problem, Cooperative Search methods add a communication mechanism to the IW strategy, in order to share information between solver instances while the search is running. Sharing information can improve the probability to get a solution faster than a parallel isolated search. However, all previous experiments indicate that it is very hard to get better performance than IW [16,22].[1] Clearly, this may be explained by the overhead incurred in performing communications, but also by

---

[1] Sometimes it even degrades performance!

the uncertainty of the benefits stemming from abandoning the current state in favor of another, heuristics-based information which may or may not lead to a solution.

In this work we propose a parametric cooperative local search framework aimed at increasing the performance of parallel implementations based on the Independent Multi-Walks strategy. This framework allows the programmer to define, for each specific problem, a custom trade-off between *intensification* and *diversification* in the search process. Intensification directs the solver to explore deeply a promising part of the search space, while diversification helps to extend the search to different regions of the search space [13].

3.1 Framework Design

Figure 2 presents the general struc-ture of the framework. All available solver instances (*Explorers*) are grouped into *Teams*. Each team implements a mechanism to ensure intensification in the search space, swarming to the most promising neighborhood found by the team members. Simultaneously, all the teams implement a mechanism to collectively provide diversification for the search (outside the groups). The

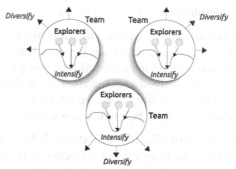

Fig. 2. Cooperative Framework Overview

expected effect is that different teams will work on different regions of the search space. Inter-team communication is needed to ensure diversification while intra-team communication is needed for intensification. This framework is oriented towards distributed architectures based on clusters of nodes: teams are mapped to nodes and explorers run on cores. For efficiency reasons it will be necessary to limit inter-node (ie. inter-team) communication.

The first parameter of the framework is the number of nodes per team (nodes_per_team), which is directly related to the trade-off between intensification and diversification. This parameter takes values from 1 to the maximum number of nodes (frequently linked to maximum number of available cores for the program in IW). When nodes_per_team is equal to 1, the framework coincides with the IW strategy, it is expected that each 1-node team be working on a different region of the search space, without intensification. When the nodes_per_team is equal to the maximum number of nodes, the framework has the maximum level of intensification, but there is no diversification at all (only 1 team available). Both diversification and intensification mechanisms are based on the use of communication between nodes. We will explain the precise role of each one in the following section.

Although we presented the framework with Local Search, it is clearly applicable to other metaheuristics as well, such as Simulated Annealing, Genetic Algorithms, Tabu Search, neighboring search, Swarm Optimization, or Ant-Colony optimization. It is also possible to combine different algorithms in a portfolio approach. For instance a team could implement a local search method, a second team could use a pure tabu search heuristics and another team could try to find a solution using a genetic algorithm. Inside a same team it is also possible to use different versions of a given metaheuristics (e.g. with different values for control parameters).

3.2   Ensuring Diversification

To provide diversification we propose a communication mechanism between teams. The teams share information to compute their current *distance* to other teams (distance between current configurations in the search space). Thus, if two groups are too close, a corrective action is executed. The parameters of this mechanism are defined as follows.

Inter-team Communication Topology: This parameter defines the way in which the communications between teams is done. For instance, in the *All-to-All* Scheme each team shares information with every other team; in the *Ring* Scheme each team only shares information with the "adjacent" teams, i.e. the previous and the next teams (e.g., team 5 only communicates with teams 4 and 6). In the *Random* scheme two teams are selected randomly to communicate each other.

Inter-team Communication Interval: This parameter indicates how frequently the communication between teams occurs. One possible approach is to measure the communication interval in terms of number of iterations elapsed in the main loop of the algorithm.

Distance Function: this function is used to check the closeness of two teams (in order to detect if they are exploring a similar region of the search space). For this, the teams compare their current configurations using the distance function. A simple function can count the number of different values in both configurations (i.e. vectors). But, depending on the problem, it is possible to elaborate more complex functions (e.g., taking into account the values and/or the indexes in the vector, using weighted sums...) When a computed distance is lower than the *minimum_permissible_distance* parameter, the two teams are declared too close.

Corrective Action: this parameter controls what to do when two teams are too close. In that case one team must correct its trajectory (this can be the "worst" one, i.e. the team whose configuration's cost is the higher). As possible corrective action the team's head node can decide to update its internal state, e.g., clearing its Elite Pool (see below). It can also restart a percentage of the team's explorers, in order to force them to explore a different portion of the search space.

3.3   Ensuring Intensification

We provide intensification by means of a communication mechanism. Here also, it is possible to have different communication topologies between the explorers of a team. In our framework, we select a topology in which each node communicates with a single other node, thereby constituting a *team* (see Figure 3).

The team is composed of one *head node* and n *explorer nodes*. Explorer nodes implement a solver instance of a Local Search method. Each Explorer node periodically reports to the head node, conveying some practical information about its search process (e.g., its current configuration, the associated cost, the number of iterations reached, the number of local minimum reached, etc.). The *head node* then pro-

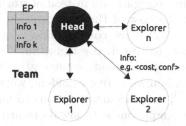

Fig. 3. Structure of a Team

cesses the messages and makes decisions to ensure the intensification in the search[2]. Moreover, the head node stores the configurations with the best costs in its Elite Pool (EP) and provides it on demand to Explorers. The head node can also decide to do some actions on its elite pool (e.g., combining configurations to create a new one, similarly to what is done in genetic algorithms).

Explorer Node. Explorer nodes periodically ask the head node for an elite configuration from the EP. If the cost of the new EP configuration is lower than its current configuration, the worker node switches to the EP one. Thus the nodes in a group intensify the search process, progressing towards the most promising neighbor found by their group. This is mainly intensification, but if an explorer node luckily finds a better configuration, then the whole team moves to this new promising neighborhood of the search space. The team is not bound to any specific region. The parameters of this mechanism are as follows:

- Report interval: The report interval parameter indicates how frequently the explorer nodes communicate information to the head node.
- Update interval: The update interval parameter indicates how frequently the explorer nodes try to obtain a new configuration from the EP in the head node.

Head Node. The head node receives and processes all the information from the explorer nodes in the team. Because of this, it has a more global vision about what is happening in the team and it can make decisions based in the comparative performance of the explorers nodes.

The main duty of the head node is to provide an *intensification* mechanism for the explorer nodes, resorting to the EP. The EP has different tuning parameters that must be defined at design time: First, the *size of the pool* which is the maximum number of configurations that the EP can store. Second, the *entry policy*, which defines the rules to accept or reject the incoming configurations. Finally the *request policy*, which defines which configuration is actually delivered to an explorer node when it makes a request.

---

[2] Recall, the head node also ensures the diversification by inter-team communication as explained above.

One possible entry policy for this framework is described below. When a message from the explorer node is received by the head node, the algorithm discards the configuration instead of storing it in the EP, in the following situations: (1) if the cost of the configuration is greater than the current worst cost in the EP, (2) if the configuration is already stored in the EP. If the configuration is not discarded, the algorithm then looks for a free slot in the EP. If there is one, the incoming configuration is stored. If not, the algorithm selects a victim configuration (e.g., random, worst, etc.) to be replaced by the new one.

There are many options to implement the request policy in the head node. A simple one is to always return the best configuration in the EP, or any (randomly chosen) configuration of the EP. Also, it is possible to implement more sophisticated algorithms. For instance, a mechanism where the probability of a configuration being selected from the EP is tied to its cost. We may even create a mutation mechanism on the EP, inspired in genetic algorithms [25,5], aspiring to improve the quality of the current configurations.

Although the main function of the head node is to provide intensification within the team, there exist many smart activities that the head node can carry out based in the collected information. For example, it can improve the efficiency of all the nodes in the team by comparing its performance and take corrective decisions, even before an event happens in the explorer node. Also, path relinking techniques [11] can be applied when different local minima have been detected.

4   An X10 Implementation

In order to verify the performance of our cooperative search strategy, we implemented a prototype of the framework using the Adaptive Search method, written in the X10 programming language.

X10 [20] is a general-purpose language developed at IBM, which provides a PGAS variant, Asynchronous PGAS (APGAS), which makes it more flexible and usable even in non-HPC platforms [19]. With this model, X10 supports different levels of concurrency with simple language constructs.

There are two main abstractions in the X10 model: *places* and *activities*. A *place* is the abstraction of a virtual shared-memory process, it has a coherent portion of the address space. The X10 construct for creating a place in X10 is the at operation, and is commonly used to create a place for each physical processing unit. An *activity* is the mechanism which abstracts the single threads that perform computation within a place. Multiple activities may be simultaneously active in one place.

Regarding communication, an *activity* may reference objects in other *places*. However, an activity may synchronously access data items only in the place in which it is running. If it becomes necessary to read or modify an object at some other place, the place-shifting operation at may be used. For more explicit communication, the GlobalRef construct allows cross-place references. GlobalRef includes information on the place where an object resides, therefore an activity may locally access the object by *moving* to the corresponding place.

A detailed examination of X10, including tutorials, language specification and examples may be found at http://x10-lang.org/.

To implement our framework in X10, we mapped each explorer node to one X10 place, using a solver instance of the Adaptive Search method as in the Independent Multi-Walks strategy. In this implementation, the head nodes also act as explorer nodes in their "spare" time.

The parameter nodes_per_team is passed to the main program as a external value. The program reads this value and creates all the instances of the solver together with the necessary references to perform the communication between nodes within a team and between the different teams in the execution.

We used the construct GlobalRef to implement communication in X10. Every *head node* reference is passed to the relevant *explorer nodes* of the team, and to the other head nodes in the program. The main loop of the solver has code to trigger all the events in the framework: *Inter-Team Communication event* (between teams), *Report event* (between the explorer nodes and head node into a team) and *Update event* (explorer nodes request a new configuration from the head node).

In the initial implementation, we opted for each explorer node only communicating its current configuration and cost pair ⟨configuration, cost⟩ to its head node. In the request event, we chose to send a random configuration from the EP to the explorer nodes. For simplicity, this first implementation does not communicate between head nodes of different teams, so diversification is only granted by the randomness of the initial point and the different seeds in each node.

5   Results and Analysis

In this section we compare our X10 implementation[3] of our framework to the independent Multi-Walks version in order to see the gain in terms of speed-ups. For this experiment, we the set of problems presented in section 2.

We used different values for parameter *nodes_per_team*: 2, 4, 8 and 16. The *Report Interval* and the *Update Interval* parameters were set to 100 iterations, finally we tried values from 1 to 4 as the size of the EP. We only retained the results for the best performing parameters. For all cases, we ran 100 samples and averaged the times.

Table 1 compares the Independent Multi-Walks implementation (IW) to our Cooperative Multi-Walks implementation (CW) for each of the problems (LNP, AIP, MSP and CAP). For each problem, a pair of rows presents the speed-up factor of the cooperative strategy CW w.r.t. the independent strategy IW (the best entry in each column is in **bold** fold).

In figure 4 we visualize same results, in a more directly perceptible form. The speed-ups obtained with IW (dotted line) and CW (continuous line) clearly show that in most cases, we are getting closer to a "ideal" speedup. It is worth noticing that AIP is unaffected by the cooperative solver when using a small numbers of cores and worse when using 24 or 32 cores. However, For the LNP, MSP and

---

[3] Source at https://github.com/dannymrock/CSP-X10.git, branch teamplaces.

Table 1. Timings and speed-ups for IW and CW on a distributed system

Problem	time (s) 1 core	Strategy	Speed-up with k cores				time (s) 32 cores
			8	16	24	32	
AIP-400	280	IW	7.1	13.1	15.3	**19.5**	**14.3**
		CW	**7.3**	**15.3**	**17.3**	19.2	14.6
speedup gain			3.5 %	17 %	13 %	-1.6 %	
LNP-500	19.1	IW	2.1	2.0	2.2	2.1	8.95
		CW	**2.5**	**3.1**	**3.4**	**3.4**	**5.7**
speedup gain			22 %	56 %	54 %	57 %	
MSP-200	274	IW	6.1	8.6	10.5	11.5	23.9
		CW	**8.5**	**14.6**	**15.7**	**18.9**	**14.6**
speedup gain			39 %	69 %	50 %	64 %	
CAP-20	992	IW	7.6	13.2	21.5	28.2	35.2
		CW	**8.9**	**16.8**	**27.6**	**32.2**	**30.8**
speedup gain			18 %	27 %	28 %	15 %	

CAP the results clearly show that the cooperative search significantly improves on the performance of the Independent Multi-Walks approach. For instance, in CAP the cooperative strategy actually reaches super linear speed-ups over the entire range of cores (speed-up of 32.2 with 32 cores). The best gain reaches 69% in the MSP.

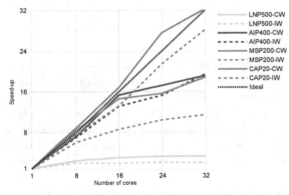

This experiments we carried out show that our cooperative framework can improve the time to find a solution for challenging instances in three of four benchmarks. It is clear that the overall performance of our cooperative teams strategy is better than the Independent Multi-Walks implementation. The main source of improvement can be attributed to the search intensification achieved

Fig. 4. Speed-Up CW vs IW

within each team. Intensification ensures that the search always stays in the best neighborhood found by the team. However, diversification is also necessary to ensure the entire set of cores does not get stuck in a local minimum.

6   Conclusion and Further Work

Following up on previous work on parallel implementations, in this paper we are concerned with the design of a general cooperative framework for parallel execution of local search algorithms, enabling a wide range of experimentation.

We decided to work with X10 as the implementation language, because it abstracts over many interesting parallel architectures while retaining a general-purpose stance. The general organization of the proposed framework entails structuring the workers as teams, each with the mission of intensifying the search in a particular region of the search space. The teams are then expected to communicate among themselves to promote search diversification. The concepts and entities involved are all subject to parametric control (e.g., trade-off between intensification and diversification, the team communication topology,...).

The initial experimentation which we carried out with an early prototype already proved to outperform the independent Multi-Walks parallel approach, even with very incomplete parameter tuning. We find these results very encouraging, suggesting that we proceed along this line of work, by defining new organizational and operational parameters as well as extending the experimentation with the ones already introduced.

This being only a preliminary work, and looking forward, we will continue to explore different communication patterns and topologies. The framework we presented relies on Local Search but it is not limited to it. We therefore plan on experimenting with other meta-heuristics or a portfolio search scheme. This is also made convenient by X10's object-oriented setting. It is also important to figure out why problems such as the All-Interval Series (AIP) do not benefit from cooperation among solvers: is it intrinsic to a certain class of problems? Which problems? Can we improve performance with different settings of the framework parameters?

Acknowledgments. The authors wish to acknowledge the Computer Science Department of UNL (Lisbon) for granting us access to its computing resources.

References

1. Arbelaez, A., Codognet, P.: Massively Parallel Local Search for SAT. In: 2012 IEEE 24th International Conference on Tools with Artificial Intelligence (ICTAI), Athens, pp. 57–64. IEEE (November 2012)
2. Caniou, Y., Codognet, P., Diaz, D., Abreu, S.: Experiments in Parallel Constraint-Based Local Search. In: Hao, J.-K., Merz, P. (eds.) EvoCOP 2011. LNCS, vol. 6622, pp. 96–107. Springer, Heidelberg (2011)
3. Codognet, P., Díaz, D.: Yet another local search method for constraint solving. In: Steinhöfel, K. (ed.) SAGA 2001. LNCS, vol. 2264, pp. 73–90. Springer, Heidelberg (2001)
4. Codognet, P., Diaz, D.: An Efficient Library for Solving CSP with Local Search. In: 5th International Conference on Metaheuristics, Kyoto, Japan, pp. 1–6 (2003)
5. Cortes, O.A.C., da Silva, J.C.: A Local Search Algorithm Based on Clonal Selection and Genetic Mutation for Global Optimization. In: 2010 Eleventh Brazilian Symposium on Neural Networks, pp. 241–246. IEEE (2010)
6. Crainic, T.G., Gendreau, M., Hansen, P., Mladenovic, N.: Cooperative parallel variable neighborhood search for the p-median. Journal of Heuristics 10(3), 293–314 (2004)

7. Diaz, D., Abreu, S., Codognet, P.: Targeting the Cell Broadband Engine for constraint-based local search. Concurrency and Computation: Practice and Experience (CCP&E) 24(6), 647–660 (2011)
8. Diaz, D., Richoux, F., Caniou, Y., Codognet, P., Abreu, S.: Parallel Local Search for the Costas Array Problem. In: Parallel Computing and Optimization, PCO 2012, Shanghai, China. IEEE (May 2012)
9. Gent, I.P., Walsh, T.: CSPLib: a benchmark library for constraints. Technical report (1999)
10. Glover, F., Laguna, M.: Tabu Search. Kluwer Academic Publishers (July 1997)
11. Glover, F., Laguna, M., Martí, R.: Fundamentals of Scatter Search and Path Relinking. Control and Cybernetics 29(3), 653–684 (2000)
12. Gonzalez, T. (ed.): Handbook of Approximation Algorithms and Metaheuristics. Chapman and Hall / CRC (2007)
13. Hoos, H., Stützle, T.: Stochastic Local Search: Foundations and Applications. Morgan Kaufmann / Elsevier (2004)
14. Kadioglu, S., Sellmann, M.: Dialectic Search. In: Gent, I.P. (ed.) CP 2009. LNCS, vol. 5732, pp. 486–500. Springer, Heidelberg (2009)
15. Machado, R., Abreu, S., Diaz, D.: Parallel local search: Experiments with a pgas-based programming model. CoRR, abs/1301.7699 (2013), Proceedings of PADL 2013, Rome, Italy
16. Machado, R., Abreu, S., Diaz, D.: Parallel Performance of Declarative Programming Using a PGAS Model ((forthcoming)). In: Sagonas, K. (ed.) PADL 2013. LNCS, vol. 7752, pp. 244–260. Springer, Heidelberg (2013)
17. Munera, D., Diaz, D., Abreu, S.: Towards Parallel Constraint-Based Local Search with the X10 Language. In: 20th International Conference on Applications of Declarative Programming and Knowledge Management (INAP), Kiel, Germany (2013)
18. Pascal, V.H., Laurent, M.: Constraint-Based Local Search. The MIT Press (2005)
19. Saraswat, V., Almasi, G., Bikshandi, G., Cascaval, C., Cunningham, D., Grove, D., Kodali, S., Peshansky, I., Tardieu, O.: The Asynchronous Partitioned Global Address Space Model. In: The First Workshop on Advances in Message Passing, Toronto, Canada, pp. 1–8 (2010)
20. Saraswat, V., Bloom, B., Peshansky, I., Tardieu, O., Grove, D.: X10 language specification - Version 2.3. Technical report (2012)
21. Schulte, C., Tack, G., Lagerkvist, M.: Modeling and Programming with Gecode (2013)
22. Toulouse, M., Crainic, T., Gendreau, M.: Communication Issues in Designing Cooperative Multi-Thread Parallel Searches. In: Meta-Heuristics: Theory & Applications, pp. 501–522. Kluwer Academic Publishers, Norwell (1995)
23. Truchet, C., Richoux, F., Codognet, P.: Prediction of parallel speed-ups for las vegas algorithms. In: 43rd International Conference on Parallel Processing, ICPP 2013. IEEE Press (October 2013)
24. Verhoeven, M.G.A., Aarts, E.H.L.: Parallel local search. Journal of Heuristics 1(1), 43–65 (1995)
25. Zhang, Q., Sun, J.: Iterated Local Search with Guided Mutation. In: IEEE International Conference on Evolutionary Computation, pp. 924–929. IEEE (2006)

A Survey of Meta-heuristics
Used for Computing Maximin Latin Hypercube

Arpad Rimmel[1] and Fabien Teytaud[2]

[1] Supélec E3S, France
arpad.rimmel@supelec.fr
[2] Univ. Lille Nord de France, France
teytaud@lisic.univ-littoral.fr

Abstract. Finding maximin latin hypercube is a discrete optimization problem believed to be *NP*-hard. In this paper, we compare different meta-heuristics used to tackle this problem: genetic algorithm, simulated annealing and iterated local search. We also measure the importance of the choice of the mutation operator and the evaluation function. All the experiments are done using a fixed number of evaluations to allow future comparisons. Simulated annealing is the algorithm that performed the best. By using it, we obtained new highscores for a very large number of latin hypercubes.

1 Introduction

In order to realize a mathematical model of a complex system with D parameters, a fixed number s of experiments are made. The values of the parameters for each experiment can be seen as a point in a D-dimensional space. The choice of the s points is crucial because it will directly impact the quality of the model. A classical requirement for the points is to respect two criteria:

- they must be *evenly spread*. This guaranties a good exploration of the parameter space.
- they must be *non-collapsing*. This ensures the fact that even if a parameter is useless, two different experiments will not give the same result.

One way to choose the points while validating those criteria is to use maximin latin hypercube. A latin hypercube in dimension D and of size s is a set of s points in $[1..D]^s$ such that on each dimension, every point has a different value. Therefore, the points in a latin hypercube are *non-collapsing*. Latin hypercube have been introduced in [12]. An example is given on figure 1.

Let d_{min} be the minimal distance of all the distances between each pair of points. The maximin latin hypercube is the latin hypercube with the largest d_{min}. The points of a maximin latin hypercube are therefore *evenly spread*. The maximin criteria has been introduced in [9]. Several distance functions can be used. In this paper, we focus on the distance function l_2. To reduce clutter, we use the square of the distance.

The points that determine the minimal distance are called *critical points*.

C. Blum and G. Ochoa (Eds.): EvoCOP 2014, LNCS 8600, pp. 25–36, 2014.

Fig. 1. Example of latin hypercubes of size 4 and dimension 2. left: $d_{min} = 2$. right: $d_{min} = 5$.

The complexity of finding a maximin latin hypercube depends on the distance function and on the dimension. For dimension 2 and both distance function l_1 and l_∞, a polynomial algorithm giving the maximin latin hypercube has been shown in [14]. For other dimensions and other distance functions, the complexity is not known but believed to be *NP*-hard.

Several meta-heuristics have been used in the literature to tackle this problem. The most classical ones are Genetic Algorithms (GA) [10], Simulated Annealing (SA) [13] and Iterated Local Search (ILS) [4]. Those three algorithms will be the focus of this paper and will be detailed later. We can also cite:

- Periodic Design (Perm) [1]. This algorithm uses periodic structures to generate latin hypercube. It gives good results for dimension 2 and correct results for dimension 3 but after that, the algorithm is outperformed by the others. For this reason and the fact that the principle of the algorithm is really different from the other, it will not be used in following comparisons.
- Enhanced Stochastic Evolutionary algorithm (ESE) [8]. This algorithm is based on a combination of GA and SA. Due to its similarity with SA, we do not use this algorithm in our experiments. However, we compare the highscores we find in this paper with the one obtained in [7].

In this paper, we present a comparison of the efficiency of GA, SA and ILS for a fixed number of evaluations. We first describe the principle of the algorithms. We then presents different mutation operators and several evaluation functions described in the literature. Finally, we give results on the efficiency of each algorithm with the different mutation operators and evaluation functions. The best combination allows us to obtain better highscores than those found in the literature for a large number of dimensions and sizes.

2   Algorithm Descriptions

In this section, we will briefly describe each algorithm. For further information, references are given in each section.

2.1   Genetic Algorithms

Genetic algorithms [5],[3] belong to the family of Evolutionary Algorithms.They are mainly used with a discrete search space, meaning they are used to address combinatorial optimization problems. Genetic Algorithms (GA) are techniques inspired by natural evolution, with in particular these following steps : (i) inheritance, (ii) mutation, (iii) crossover and (iv) selection. The principle is to have a population of candidate solutions evolving. Under the pressure selection, the population will converge toward better solutions. Algorithm 1 illustrates this method. λ denotes the population size, i.e. the number of individuals in the population.

Algorithm 1. Genetic Algorithm

Generate the initial population
Evaluate all individuals of the population
while the stopping condition not reached **do**
    for i from 1 to λ **do**
        parent1 $\leftarrow$ `parentalSelection`(parentPopulation)
        if crossover probability is satisfied **then**
            parent2 $\leftarrow$ `parentalSelection`(parentPopulation)
            offspringPopulation[i] $\leftarrow$ `crossover`(parent1,parent2)
        else
            offspringPopulation[i] $\leftarrow$ parent1
        end if
        if mutation probability is satisfied **then**
            offspringPop[i] $\leftarrow$ `mutate`(offspringPop[i])
        end if
    end for
    Evaluate all individuals of offspringPopulation
    parentPopulation $\leftarrow$ `survivalSelection`(offspringPopulation, parentPopulation)
end while

2.2 Simulated Annealing

Simulated Annealing [2] is a global optimization algorithm, generally used with a discrete search space. The principle takes inspiration from the annealing principle in metallurgy. This technique consists in two phases : (i) heating and (ii) controlled cooling. In optimization, the method is to evolve a candidate solution thanks to a mutation. According to a certain probability a worse generated solution can be accepted. The idea behind this is to not get stuck in a local optimum. The controlled cooling represents this acceptance probability. Algorithm 2 represents this method.

2.3 Iterated Local Search

Iterated Local Search (ILS) [11] is based on a local search algorithm. An improvement over a single run of a local search is to launch it several times from a different starting point. A classical way of doing this is by choosing a new random starting point, the resulting algorithm is called the multistart approach

---

Algorithm 2. Simulated Annealing

Generate the initial solution s
Evaluate its fitness e $\leftarrow$ f(s)
$s_{best} \leftarrow s$
$e_{best} \leftarrow e$
while the stopping condition not reached **do**
    T $\leftarrow$ **temperature**(iteration)
    $s_{new} \leftarrow$ **mutate**(s)
    $e_{new} \leftarrow$ f(s_{new})
    if acceptance probability is satisfied **or** e_{new} **better than** e **then**
        $s \leftarrow s_{new}$
        $e \leftarrow e_{new}$
    end if
    if e_{new} **better than** e_{best} **then**
        $s_{best} \leftarrow s_{new}$
        $e_{best} \leftarrow e_{new}$
    end if
    increment iteration
end while

---

or random restart. The principle of ILS is to choose the new starting point by a perturbation of the last local minimum found instead.

The algorithm 3 is used for the local search. The initial solution is generated with a uniform distribution. The perturbation consists in a rotation of the values of several points on a given dimension.

In algorithm 3, the condition $*$ about the critical point is not necessary. Both possibilities will be studied in the following.

3   Mutations

In [6] 4 different neighborhoods are proposed. We have considered these 4 neighborhoods for our mutations and called them $m1$ to $m4$. The main principle of all these mutations is to change two points of the hypercube.

In the first one, the first point is chosen uniformly among all the critical points. The second point is chosen uniformly among all remaining points. For these two points, a random number of coordinates are changed.

In the second one, the two points are selected in the same way, but only one coordinate is changed. The coordinate is chosen randomly.

In the third one, the two points are selected in the same way and only one coordinate is changed. All the coordinates are tried and the one which gives the best d_{min} is selected. This modification add DIM evaluation to the algorithm, those evaluations are taken into account in the experiments.

In the last mutation, the two points are chosen uniformly among all points and the number of coordinates is randomly chosen.

Algorithm 3. Local Search of the ILS algorithm

$X = InitialSolution$
while there is an improvement **do**
    for i from 1 to size **do**
        for j from 1 to size **do**
            if i != j AND (i or j is a critical point)* **then**
                for k from 1 to DIM **do**
                    $X' = X$
                    in X' switch value of the points i and j on dimension k
                    if $eval(X') > eval(X)$ **then**
                        $X = X'$
                    end if
                end for
            end if
        end for
    end for
end while
return X

---

4 Evaluation Functions

The natural evaluation function for a latin hypercube is the d_{min} function as this is the function used to determine if the latin hypercube is maximin. An other evaluation function ϕ_p has been proposed in [13]. It has the advantage of differentiating situations that would have a similar d_{min} value by using *all* the distances between points. This function is defined as follow:

- let $D_1(X), D_2(X), ..., D_R(X)$ be the set of all the distances between two points of X ordered in increasing order. R is the number of different distances in X.
- let $J_i(X)$ be the number of occurrences of $D_i(X)$.

$$\phi_p(X) = \left(\sum_{r=1}^{R} \frac{J_r(X)}{(D_r(X))^p} \right)^{\frac{1}{p}}, p \text{ is a parameter.}$$

5 Experiments

In this section, we make several experiments with the most efficient algorithms for the latin hypercube problem: *Genetic Algorithm (GA), Simulated Annealing (SA)* and *Iterated Local Search (ILS)*. We first determine which value to use for each algorithm's parameter. Then, we evaluate the effect of the choice of the mutation function. Finally, we determine the effect of the evaluation function.

For the comparison between the algorithms to be fair and not implementation-dependant, we chose to fix the number of time the evaluation of a latin hypercube has to be computed. Unless it is stated otherwise, the number of evaluations used for the experiment is $100000 * DIM$. We checked that the execution time depends only on the number of evaluations.

As testing every combination of dimension and size would be to time consuming, we chose arbitrarily three set of values that will be used for the experiments: (DIM 4 / SIZE 25), (DIM 9 / SIZE 10) and (DIM 8 / SIZE 20).

5.1 Effect of Algorithm Parameters

In this section, we experiment with different parameter values for each algorithm in order to determine which parameter has a high influence on the performance of the algorithm. It will also be used to fix the parameters for the next sections. For those experiments, the mutation operator $m3$ and the evaluation ϕ_{20} are used.

Genetic Algorithm. The main parameters of the genetic algorithm are λ and Gen.

Table 1. Effect of the modification of the ratio λ/Gen for GA

LatinHyperCube	$\lambda100, Gen1000$	$\lambda200, Gen500$	$\lambda500, Gen200$
4/25	158.2±0.3	162.1 ± 0.2	155.9 ± 0.2
9/10	151.3±0.2	153.0 ± 0.1	153.9 ± 0.1
8/20	411.8±0.3	413.0 ± 0.2	401.6 ± 0.2

We see in table 1 that the parameters have a moderate impact on the performance. The ratio λ/Gen that gives the best performance is 2/5. This ratio will be used in the rest of the paper.

Table 2. Performance of SA for different initial and final probabilities

(DIM/SIZE)	0.5/0.1	0.3/0.1	0.5/0.01	0.3/0.01	0.1/0.01	0.05/0.01
4/25	162.64±0.1	171.97±0.1	170.24±0.09	175.24±0.07	175.87±0.08	174.9±0.1
9/10	154.33±0.02	155.35±0.02	155.15±0.02	155.34±0.02	154.86±0.04	154.07±0.07
8/20	418.71±0.1	427.88±0.1	429.64±0.09	434.02±0.08	435.33±0.07	434.84±0.08

Simulated Annealing. We see in table 2 that the value of the initial and final probabilities are very important parameters for the performance of the algorithm. We chose 0.3 for the starting probability and 0.01 for the final probability in the following experiments.

On the contrary, we see in table 3 that the ratio between the number of cycles and the number of trials is insignificant for the performance. We use a ratio of 10/1 for the rest of this paper.

Table 3. Comparison with mutation3 and eval_phi with simulated annealing

LatinHyperCube (DIM/SIZE)	# cycles	# of trials per cycles	avg scores
4/25	1000	100	175.24±0.07
	500	200	175.31± 0.07
	200	500	175.32±0.07
	100	1000	175.2±0.07
9/10	1000	100	155.34±0.02
	500	200	155.34±0.02
	200	500	155.3±0.02
	100	1000	155.34±0.02
8/20	1000	100	434.02±0.08
	500	200	434.96±0.07
	200	500	433.89±0.07
	100	1000	434.08±0.07

Iterated Local Search. Iterated Local Search does not have many parameters. We will however compare different version of the algorithm.

- v1: in the local search, all the points are considered
- v2: in the local search, only the critical points are considered.
- v3: multistart version of the algorithm v1.

Table 4. Performance of ILS for different versions of the algorithm

LatinHyperCube	ISL v1	ILS v2	ILS v3
4/25	162.0±0.2	158.5± 0.3	159.7±0.2
9/10	153.3±0.1	152.7± 0.1	151.0±0.1
8/20	412.7±0.3	408.0± 0.3	409.5±0.2

The different version have similar performance with a statistically significant advantage for the v1 version. This version will be used in the following sections.

5.2   Effect of the Mutations

We compare the performance of the genetic algorithm and the simulated annealing algorithm with the different mutations described in section 2. The results are given in table 2.

The choice of the mutation operator has a huge impact on the performance of both algorithms. This seems like a good direction for future improvements. In the rest of the paper, we will use the mutation 3.

5.3   Effect of the Evaluation Function

In this section, we compare the performance of the different algorithms when optimizing using the d_{min} function of using the ϕ_p function. The score is always

Table 5. Comparison of the effect of different mutation operators on the performance
of the algorithms

LatinHyperCube	mutation	GA	SA
4/25	m1	156.6±0.2	173.82±0.12
	m2	156.4±0.3	173.64±0.12
	m3	162.0±0.2	175.22±0.07
	m4	147.2±0.3	173.69±0.09
9/10	m1	149.7±0.2	155.77±0.02
	m2	152.0±0.1	155.88±0.02
	m3	152.9±0.1	155.34±0.02
	m4	147.7±0.2	155.68±0.02
8/20	m1	399.4±0.3	429.79±0.13
	m2	406.1±0.3	425.8±0.13
	m3	412.8±0.2	433.92±0.07
	m4	388.8±0.4	424.59±0.13

the d_{min} value, even when optimizing with ϕ_p. The results are presented in
table 6.

Optimizing according to the ϕ_p function improves greatly the results for all 3
algorithms. The value of the parameter p seems to have not so much impact as
long as it is greater than 1. In the following, we will use ϕ_{10} as our evaluation
function.

5.4   Scalability of the Algorithms

The scalability of an algorithm represents the robustness of the algorithm in
front of the number of evaluations. For instance, if an algorithm performs better
with a large number of evaluations, it should be a good choice if the evaluations
can be parallelized.

We launched each algorithm with its best configuration with different number
of evaluations and measured the best d_{min} obtained. The results are given on
figure 2.

SA performs better than the two other algorithms in every case. Furthermore,
there is no algorithm that improves faster than the others when the number of
simulations increases. So SA should stay the best even with larger number of
simulations.

6   HighScores

A highscore is the best d_{min} value obtained for a particular $SIZE$ and DIM. It
corresponds to the best known lower bond for the maximin value. The previous
highscores were obtained with the algorithms Iterated Local Search (see [4]) and
ESE (see [7]). According to our comparison, the simulated annealing algorithm
with the mutation operator m3 and the evaluation function ϕ_{10} performs better.
We launched this algorithm on latin hypercube of dimension 3 to 10 and of

Table 6. Comparison of the effect of different evaluation function on the performance of the algorithms

LatinHyperCube	Evaluation Function	GA	SA	ILS
4/25	d_{min}	142.4±0.3	155.75±0.18	118.2±0.4
	ϕ_1	153.6±0.2	166.07±0.06	149.9±0.3
	ϕ_5	167.4±0.2	176.47±0.06	165.7±0.2
	ϕ_{10}	165.8±0.2	176.47±0.06	165.5±0.2
	ϕ_{20}	161.8±0.2	175.26±0.07	162.0±0.2
9/10	d_{min}	144.5±0.2	150.00±0.10	138.3±0.2
	ϕ_1	153.3±0.1	155.58±0.04	153.8±0.1
	ϕ_5	153.4±0.1	155.68±0.02	153.9±0.1
	ϕ_{10}	153.5±0.1	155.64±0.02	153.7±0.1
	ϕ_{20}	153.0±0.1	155.29±0.02	153.3±0.1
8/20	d_{min}	381.5±0.4	406.18±0.21	332.6±0.7
	ϕ_1	396.5±0.3	415.16±0.11	376.2±0.5
	ϕ_5	410.4±0.2	430.68±0.09	404.1±0.3
	ϕ_{10}	414.9±0.2	434.29±0.07	413.1±0.2
	ϕ_{20}	413.2±0.2	434.06±0.08	412.7±0.2

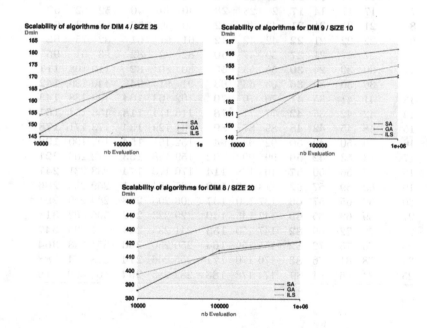

Fig. 2. Effect of the increase of the number of evaluations. We plot the minimal distance as a function of the number of evaluations. The higher the better.

size 2 to 25 with 10 000 000 evaluations. The tables 7 and 8 show the previous highscores obtained with ILS and ESE as well as the ones we get with SA. On a large majority of the latin hypercubes, we obtain similar or better result.

Table 7. Current highscores

SIZE	DIM = 3				DIM = 4			DIM = 5			DIM = 6		
	ILS	ESE	Per	SA	ILS	ESE	SA	ILS	ESE	SA	ILS	ESE	SA
2	3	3	3	**3**	4	4	**4**	5	5	**5**	6	6	**6**
3	6	6	3	**6**	7	7	**7**	8	8	**8**	12	12	**12**
4	6	6	6	**6**	12	12	**12**	14	14	**14**	20	20	**20**
5	11	11	6	**11**	15	15	**15**	24	24	**24**	27	27	**27**
6	14	14	14	**14**	22	22	**22**	32	32	**32**	40	40	**40**
7	17	17	14	**17**	28	28	**28**	40	40	**40**	52	52	**52**
8	21	21	21	**21**	42	42	**42**	50	50	**50**	66	63	**66**
9	22	22	21	**22**	42	42	**42**	61	61	**61**	82	75	**82**
10	27	27	21	**27**	50	47	**50**	82	82	**82**	93	91	**95**
11	30	30	24	**30**	55	55	**55**	80	80	**82**	110	108	**111**
12	36	36	30	**36**	63	63	**63**	91	91	**94**	140	136	**142**
13	41	41	35	**41**	70	70	**70**	102	103	**104**	139	138	**143**
14	42	42	35	**42**	77	77	**78**	116	114	**118**	156	154	**161**
15	48	45	42	**48**	89	87	**89**	131	129	**134**	173	171	**178**
16	50	50	42	**50**	92	93	**94**	152	151	**154**	192	190	**196**
17	53	53	42	**56**	99	99	**102**	159	158	**163**	212	208	**221**
18	57	56	50	**57**	108	108	**114**	170	170	**174**	238	231	**247**
19	62	59	57	**62**	118	119	**122**	187	184	**191**	259	256	**266**
20	66	65	57	**66**	137	130	**137**	206	206	**208**	284	279	**291**
21	69	68	65	**69**	147	145	**149**	229	223	**232**	306	302	**314**
22	76	72	69	**82**	147	150	**153**	241	235	**244**	331	325	**347**
23	76	75	72	**82**	155	159	**164**	250	250	**260**	351	348	**364**
24	78	81	76	**83**	170	170	**171**	265	266	**274**	378	374	**388**
25	84	86	91	**89**	174	178	**183**	282	285	**294**	408	400	**419**

Table 8. Current highscores

SIZE	DIM = 7			DIM = 8			DIM = 9			DIM = 10		
	ILS	ESE	SA	ILS	ESE	SA	ILS	ESE	SA	ILS	ESE	SA
2	7	7	**7**	8	8	**8**	9	9	**9**	10	10	**10**
3	13	13	**13**	14	14	**14**	18	18	**18**	19	19	**19**
4	21	21	**21**	26	26	**26**	28	28	**28**	33	33	**33**
5	32	32	**32**	40	40	**40**	43	43	**43**	50	50	**50**
6	47	47	**47**	54	53	**54**	61	61	**62**	68	68	**68**
7	62	61	**62**	71	70	**72**	80	80	**81**	90	89	**91**
8	79	79	**80**	91	90	**91**	102	101	**103**	114	114	**116**
9	95	92	**95**	114	112	**114**	128	126	**128**	143	142	**144**
10	112	109	**113**	133	131	**133**	157	154	**158**	173	171	**175**
11	131	129	**132**	153	152	**157**	181	178	**184**	209	206	**211**
12	155	152	**158**	181	177	**183**	208	204	**213**	240	235	**243**
13	181	178	**184**	208	205	**213**	242	235	**246**	274	268	**279**
14	219	215	**220**	242	236	**245**	275	268	**282**	312	305	**318**
15	222	220	**228**	277	273	**280**	315	309	**321**	356	347	**363**
16	249	241	**253**	326	317	325	357	352	**364**	403	393	**411**
17	269	266	**278**	331	332	**343**	404	396	**413**	451	442	**462**
18	297	291	**304**	367	361	**374**	466	451	**469**	505	496	**515**
19	323	323	**335**	398	390	**408**	466	469	**491**	569	554	**576**
20	356	349	**369**	426	425	**446**	512	506	**528**	640	625	**645**
21	386	380	**397**	467	463	**482**	550	548	**570**	647	650	**672**
22	421	418	**433**	505	501	**525**	593	595	**623**	697	691	**720**
23	452	448	**466**	545	542	**566**	649	640	**667**	736	747	**774**
24	486	481	**506**	591	585	**609**	694	690	**718**	805	800	**837**
25	525	520	**536**	631	626	**657**	746	739	**769**	866	857	**893**

7 Conclusion

We presented three classical algorithms to find maximin latin hypercube: genetic algorithm, iterated local search and simulated annealing. The experiments were done with a fixed number of evaluations to allow easier future comparisons. We show that the choice of the mutation operator has a great impact on the performance of each algorithm. Using a surrogate evaluation function also significantly improve the results. We then compared the 3 algorithms and the way they scale up and find that simulated annealing performs better than the other. Finally, we used simulated annealing on longer time settings to obtain scores as good as possible for different latin hypercube. This allowed us to obtain new highscores on most of them.

References

1. Bates, S.J., Sienz, J., Toropov, V.V.: Formulation of the optimal latin hypercube design of experiments using a permutation genetic algorithm. AIAA 2011, 1–7 (2004)

2. Bohachevsky, I.O., Johnson, M.E., Stein, M.L.: Generalized simulated annealing for function optimization. Technometrics 28(3), 209–217 (1986)
3. Goldberg, D.E., Holland, J.H.: Genetic algorithms and machine learning. Machine Learning 3(2), 95–99 (1988)
4. Grosso, A., Jamali, A., Locatelli, M.: Finding maximin latin hypercube designs by iterated local search heuristics. European Journal of Operational Research 197(2), 541–547 (2009)
5. Holland, J.H.: Adaptation in natural and artificial systems: An introductory analysis with applications to biology, control, and artificial intelligence. U. Michigan Press (1975)
6. Husslage, B., Rennen, G., Van Dam, E.R., Den Hertog, D.: Space-filling Latin hypercube designs for computer experiments. Tilburg University (2006)
7. Husslage, B.G., Rennen, G., van Dam, E.R., den Hertog, D.: Space-filling latin hypercube designs for computer experiments. Optimization and Engineering 12(4), 611–630 (2011)
8. Jin, R., Chen, W., Sudjianto, A.: An efficient algorithm for constructing optimal design of computer experiments. Journal of Statistical Planning and Inference 134(1), 268–287 (2005)
9. Johnson, M.E., Moore, L.M., Ylvisaker, D.: Minimax and maximin distance designs. Journal of Statistical Planning and Inference 26(2), 131–148 (1990)
10. Liefvendahl, M., Stocki, R.: A study on algorithms for optimization of latin hypercubes. Journal of Statistical Planning and Inference 136(9), 3231–3247 (2006)
11. Lourenço, H.R., Martin, O.C., Stützle, T.: Iterated local search. International series in operations research and management science, pp. 321–354 (2003)
12. McKay, M.D., Beckman, R.J., Conover, W.J.: Comparison of three methods for selecting values of input variables in the analysis of output from a computer code. Technometrics 21(2), 239–245 (1979)
13. Morris, M.D., Mitchell, T.J.: Exploratory designs for computational experiments. Journal of Statistical Planning and Inference 43(3), 381–402 (1995)
14. Van Dam, E.R., Husslage, B., Den Hertog, D., Melissen, H.: Maximin latin hypercube designs in two dimensions. Operations Research 55(1), 158–169 (2007)

An Analysis of Parameters of irace

Leslie Pérez Cáceres, Manuel López-Ibáñez, and Thomas Stützle

IRIDIA, CoDE, Université libre de Bruxelles, Belgium
{leslie.perez.caceres,manuel.lopez-ibanez,stuetzle}@ulb.ac.be

Abstract. The irace package implements a flexible tool for the automatic configuration of algorithms. However, irace itself has specific parameters to customize the search process according to the tuning scenario. In this paper, we analyze five parameters of irace: the number of iterations, the number of instances seen before the first elimination test, the maximum number of elite configurations, the statistical test and the confidence level of the statistical test. These parameters define some key aspects of the way irace identifies good configurations. Originally, their values have been set based on rules of thumb and an intuitive understanding of the configuration process. This work aims at giving insights about the sensitivity of irace to these parameters in order to guide their setting and further improvement of irace.

1 Introduction

Algorithm configuration [5, 9] is the task of finding a setting of the categorical, ordinal, and numerical parameters of a target algorithm that exhibit good empirical performance on a class of problem instances. Currently, few tools are available for configuring algorithms automatically [1, 4, 10–12]. The irace package [12] implements an iterated racing framework for the automatic configuration of algorithms [3, 6]. irace is currently available as an R package and the details of its implementation and a tutorial on how to use it can be found in [12]. The implementation of irace is flexible, allowing the user to adjust the configuration process according to the configuration scenario at hand. As a flexible tool, irace itself has parameters. The default parameter settings of irace have been defined by rules of thumb based on intuition of how the configuration process may work [3, 6, 12]. So far, experimental analysis of these parameters have focused on a single race, studying the effect of the number of initial configurations, the particular statistical test or the confidence level [5, 7]. This paper is the first to empirically study the impact that specific settings of irace parameters have on the effectiveness of the configuration process. Section 2 describes details of irace and its default settings, Section 3 describes the experimental setup and Section 4 presents the experiments results. We conclude in Section 5.

2 The irace Procedure

Automatic algorithm configuration tools, henceforth called *configurators*, are algorithms that tackle expensive, stochastic nonlinear mixed-variable optimization

C. Blum and G. Ochoa (Eds.): EvoCOP 2014, LNCS 8600, pp. 37–48, 2014.
© Springer-Verlag Berlin Heidelberg 2014

problems. The problem tackled by a configurator is called a *configuration scenario* and it is given as a target algorithm to be configured, a set of training instances representative of the problem to be solved by the target algorithm, and the configuration budget, which is the maximum computational effort (e.g., number of runs of the target algorithm) that the configurator has available. In addition, configurators have themselves parameters that affect their search.The search of `irace` consists of a number of iterations. In each iteration, a set of candidate algorithm configurations is generated and the best configurations of the iteration are identified by racing. Within a race, configurations are tested on a sequence of problem instances and, at each step, all surviving configurations are tested on a new instance. Candidate configurations are eliminated from the race if they are found to be poor performing according to some criterion. In `irace`, this criterion is implemented by means of statistical testing. For `irace` a minimum number of iterations (N^{iter}) is defined as $N^{iter} = \lfloor 2 + \log_2(N^{param}) \rfloor$, where N^{param} is the number of parameters of the target algorithm. The user-defined budget (B, the maximum number of target algorithm runs) is distributed across the iterations by setting the budget B_i available for iteration i to

$$B_i = \frac{(B - B_i^{used})}{N^{iter} - i + 1} \tag{1}$$

where B_i^{used} is the budget that has already been used before iteration i. An iteration is stopped as soon as the budget is spent or the number of candidates in the race reaches N^{max}, defined by $N^{max} = \lfloor 2 + \log_2(N^{param}) \rfloor$. If an iteration is stopped due to this latter condition, the iteration budget may not be used completely, and thus there may be enough budget to do more than N^{iter} iterations. Each iteration starts with a set of C_i configurations:

$$C_i = \left\lfloor \frac{B_i}{\max(\mu, T^{first}) + \min(5, i)} \right\rfloor \tag{2}$$

where μ ensures a minimum number of instances seen per iteration (five by default) and T^{first} is explained below. Using this definition of C_i, the number of candidates sampled decreases with the iteration number. This was done to account for the effect that candidate configurations become more similar as the configuration process progresses and more problem instances are needed to discriminate between them. In the first iteration of `irace`, an initial set of candidate configurations may be specified, the other candidates are generated uniformly at random. In the following iterations, the set of candidates is formed by the best candidates of the previous iteration and by new candidates that are sampled around these best candidates. For numerical (integer or continuous) parameters a truncated normal distribution is used and for categorical parameters a discrete one. Ordinal parameters are treated as integers. The distributions are updated every iteration, biasing the sampling towards the best candidates found. Each race evaluates the current set of candidates on a sequence of problem instances. Candidates are discarded from the race as soon as they show statistically worse performance than the best candidate so far. In the current `irace` implementation, either the Friedman test with its associated post-test [8] or a Student t-test can be used as statistical test. The first statistical test of an iteration is performed after seeing T^{first} instances ($T^{first} = 5$, by default). The survivors are evaluated on the next instances and every T^{each} instances a statistical test is applied ($T^{each} = 1$, by default). A race finishes when at most N^{max} survivors remain in the race or the available budget B_i is exhausted. At the end of an iteration, the best candidates are selected from the survivors; these candidates are called elite candidates. The number of elite candidates in an iteration is given by $N_i^{elite} = \min\{N_i^{surv}, N^{max}\}$, where N_i^{surv} is the number of candidates that remain in the race when iteration i is finished. The selection of the elite candidates is done by ranking them (according to the sum of ranks for the Friedman test or the mean quality for t-test), and selecting from the lowest ranked. The elite candidates are then used to generate new candidates; to do so, they are selected to become parents with probability;

$$ p_z = \frac{N_i^{elite} - r_z + 1}{N_i^{elite} \cdot (N_i^{elite} + 1)/2} \tag{3} $$

where N_i^{elite} is the actual number of elite candidates in iteration i and r_z is the ranking of elite candidate z. New candidates are sampled according to the distribution associated to each parameter in the selected parent. Once C_i candidates are obtained (including the N_i^{elite} ones and newly sampled ones), a new race begins. `irace` terminates when the total budget is exhausted or when the remaining budget is not enough to perform a new iteration. Finally, the best configuration found is returned.

3   Experimental Setup

In this section, we detail the configuration scenarios used for analyzing `irace`. Each scenario has a target algorithm, a set of training and test instances and a budget. The configuration scenarios are available at the supplementary information page (`http://iridia.ulb.ac.be/supp/IridiaSupp2013-008/`).

3.1   Configuration Scenarios

ACOTSP is a software package that implements various ant colony optimization (ACO) algorithms [17] for solving the Traveling Salesman Problem (TSP). The ACOTSP scenario requires the configuration of 11 parameters of ACOTSP, three categorical, four integer and four continuous. The training set is composed of ten random Euclidean TSP instances of each of 1000, 1500, 2000, 2500 and 3000 cities; the test set has 50 instances of each of the previous sizes. All instances and their optimal solutions are available from the supplementary pages. The goal is to minimize tour length. The maximum execution time of a run of ACOTSP is set to 20 seconds and the total configuration budget to 5000 runs.

   SPEAR is a tree search solver for SAT problems [2]. The SPEAR scenario requires the configuration of 26 parameters of SPEAR, all of them categorical.

The training and the test set are composed of 302 SAT instances each, which belong to the SAT configuration benchmark "Spear-swv". The goal is to minimize mean algorithm runtime. The maximum execution time for each run of SPEAR is set to 300 seconds and the total configuration budget is 10000 runs.

MOACO is a framework of multi-objective ACO algorithms [13]. The MOACO scenario requires the configuration of 16 parameters: 11 categorical, one integer and four real. The training and the test set are composed each of 10 instances of 500, 600, 700, 800, 900, 1000 cities. The goal is to optimize the quality of the Pareto-front approximation as measured by the hypervolume quality measure [18]. The hypervolume is to be maximized, however, for consistency with the other scenarios, we plot the negative normalized hypervolume, which is to be minimized. The maximum execution time of each run of MOACO is defined by $4 \cdot (instance\_size/100)^2$. The total configuration budget is 5000 runs.

3.2   Training Set Analysis

The homogeneity of the training set with respect to algorithm performance is conjectured to have a high impact on the configuration process and possibly on the parameter settings of configurators. Homogeneity refers to the correlation between algorithm performance across instance sets: highly homogeneous instance sets maintain the same relative ranking of algorithms; highly heterogeneous instance sets lead to strongly different rankings depending on the particular problem instance to be tackled. Consequently, highly heterogeneous sets hinder the progress of the configuration process as candidate algorithm configurations may have inconsistent performance. A parametric measure of instance set homogeneity was proposed in [14], suggesting that the Friedman test statistic may be useful. As this latter statistic is not normalized and, thus, depends on the number of instances and configurations, we use the Kendall concordance coefficient (W) [15], which is a normalization of the Friedman test statistic. For each training set we generate 100 candidates uniformly distributed in the configuration parameter space. These candidates are evaluated on the instance set and Kendall's W statistic is calculated using instances as blocks and candidates as groups. The statistic of this test can be interpreted as a measure of how similar is the relative performance of candidates (that is, their ranking) across the instance set. A value close to one indicates high homogeneity, a value close to zero high heterogeneity. We performed the test using the complete instance sets and subsets grouped by instance size (ACOTSP and MOACO) or instance type (SPEAR). As shown by Table 1, the instances used in the ACOTSP and MOACO scenarios are much more homogeneous than those in the SPEAR scenario.

3.3   Experimental Setup

In the following sections, each experiment consists of 20 trials of `irace`, resulting in 20 final best configurations for each configuration scenario and each parameter setting of `irace`. For each configuration obtained, the average performance on

Table 1. Kendall's W statistic measured across 100 algorithm configurations on the training sets (*all*) and subsets grouped by size or type. The *Set* column is the set or subset of instances, the *Size* column is the number of instances in the set.

ACOTSP			SPEAR			MOACO		
Set	Size	W	Set	Size	W	Set	Size	W
all	50	0.96974	all	302	0.16017	all	60	0.99049
1000	10	0.98227	dspam	49	0.15446	500	10	0.99152
1500	10	0.98125	gzip	37	0.38442	600	10	0.99206
2000	10	0.98250	hsat	148	0.15510	700	10	0.99322
2500	10	0.98493	itox	26	0.61934	800	10	0.99256
3000	10	0.98089	winedump	17	0.29974	900	10	0.99311
			winegcc	22	0.62083	1000	10	0.99096
			xinetd	3	0.35308			

Table 2. Wilcoxon signed-rank test p-values comparing the mean performance over the test instances of configurations obtained by `irace` using $N^{iter} \in \{1, 3, \text{default}, \text{large}\}$

	default vs. large			default vs. 3			default vs. 1		
	ACOTSP	SPEAR	MOACO	ACOTSP	SPEAR	MOACO	ACOTSP	SPEAR	MOACO
F-test	0.33	0.4304	0.8695	0.7562	0.7562	0.0003948	$1.907e^{-5}$	0.7285	$1.907e^{-6}$
t-test	0.2943	0.498	0.0007076	0.7285	0.4304	$1.907e^{-6}$	0.0002098	0.5459	$1.907e^{-6}$

the test set is computed. We repeat each experiment using, as elimination test either the F-test (and its associated post-hoc tests) or the Student t-test without multiple test correction.[1] The experiments were executed on a cluster running Cluster Rocks GNU/Linux 6.0. The experiments involving the ACOTSP scenario were executed on an AMD Opteron 6128 with 8 cores of 2GHz and 16GB RAM. The ones involving the SPEAR and MOACO scenarios were executed on an AMD Opteron 6272 with 16 cores of 2.1GHz and 64GB RAM.

4   Experiments

In this section, we examine the impact of five parameters of `irace` on the performance of the final algorithm configuration found in the configuration process.

Number of Iterations. The number of iterations (N^{iter}) strongly modifies the search behavior of `irace`. With more iterations, fewer configurations are used in each iteration. The number of newly sampled configurations is also reduced as the number of elite configurations remains the same. Overall, this leads to an

---

[1] Using multiple test corrections in the Student t-test results in a search process that does not effectively eliminate poor candidates [5]. Avoiding multiple test corrections makes the process more heuristic, but proves to be effective.

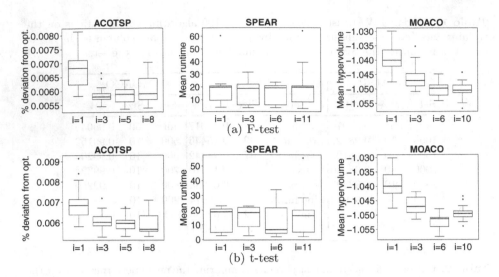

Fig. 1. Box plots of the mean performace over the test instances of 20 configurations obtained by `irace` using $N^{iter} \in \{1, 3, \text{default}, \text{large}\}$

intensification of the search by splitting the budget in short races. Less iterations, on the other hand, lead to a stronger diversification of the search. The default number of iterations of `irace` depends on the number of parameters. We increase this value to $N^{iter} = \lfloor 2 + 2 \cdot \log_2(N^{param}) \rfloor$ and we refer to this setting as "large" in the following. Additionally, we use two fixed settings: $N^{iter} = 3$ and $N^{iter} = 1$. The latter actually corresponds to a single race using configurations sampled uniformly at random [3]. In Fig. 1, we present the results of the 20 independent executions of `irace` on the three configuration scenarios and the results of the Wilcoxon test are shown in Table 2. In the SPEAR scenario, none of the differences is statistically significant, confirming the observation from the box-plots that no clear differences arise. Surprisingly, even a race based on a single random sample of configurations ($N^{iter} = 1$) obtains reasonable performance here. This is different from the MOACO and ACOTSP scenarios, where `irace` with $N^{iter} = 1$ performs significantly worse than the other settings, confirming earlier results [6]. Other differences in the ACOTSP scenario are, however, not statistically significant. In the MOACO scenario, the default setting performs significantly better than $N^{iter} = 3$, while the large setting performs significantly worse than the default only when using t-test. The results indicate that the default setting is overall reasonably robust. Nonetheless, the number of iterations has an impact on the quality of the final configurations and the adaptation of the number of iterations to the configuration scenario may be useful to improve `irace` performance.

First Elimination Test. The elimination of candidates during the race allows `irace` to focus the search around the best configurations. Here, we analyze the

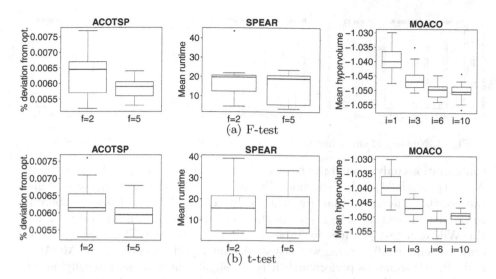

Fig. 2. Box plots of the mean performace over the test instances of 20 configurations obtained by `irace` using $T^{first} \in \{2, 5\}$

Table 3. Wilcoxon signed-rank test p-values comparing the mean performance over the test instances of configurations obtained by `irace` using $T^{first} = 2$ vs. $T^{first} = 5$

	ACOTSP	SPEAR	MOACO
F-test	0.01362	0.1231	0.1231
t-test	0.03623	0.5958	0.6477

sensibility of `irace` to the number of instances evaluated before performing the first elimination test (T^{first}). We performed experiments using the default setting of ($T^{first} = 5$) and a reduced value of $T^{first} = 2$. Reducing the value of T^{first} allows `irace` to more aggressively eliminate configurations. The budget saved in this way may be used later to sample a higher number of configurations. However, good configurations may erroneously be lost more easily. The experimental results are shown in Fig. 2. In the ACOTSP scenario a setting of $T^{first} = 2$ seems to worsen performance, while on the SPEAR and MOACO scenarios no clear differences are visible. The Wilcoxon paired test in Table 3 supports this analysis. Our hypothesis was that with a setting of $T^{first} = 2$, poor candidates are eliminated earlier and in later iterations more candidates may be sampled. In order to corroborate this hypothesis, we plot the development of the number of surviving configurations during the search process of `irace` (Fig. 3). The plots show one run of `irace` that is representative for the general behavior.

Maximum Number of Elite Configurations. The maximum number of elite configurations (N^{max}) influences the exploration / exploitation trade-off in the search process. In the extreme case of $N^{max} = 1$, `irace` samples new

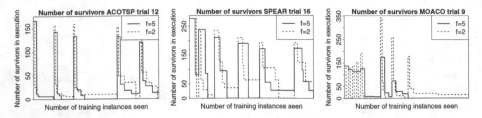

Fig. 3. Number of surviving candidates in `irace` using $T^{first} \in \{2, 5\}$ and F-test

configurations only around the best configuration found so far. A larger value of N^{max} (potentially) induces a more diverse search. In this section, we examine the possible differences that are incurred by setting $N^{max} = 1$ and compare it to the default setting. The results of these experiments are shown in Fig. 4 and the Wilcoxon test p-values in Table 4. While in the ACOTSP scenario using only one elite configuration worsens significantly performance, in the MOACO and SPEAR scenarios performance it is not significantly worse. Intensifying the search by strongly reducing the number of elite candidates does not seem to improve the performance of the final configurations in any of the configuration scenarios. These results indicate that the default setting is reasonably adequate.

Statistical Test. The main difference between the F-test (plus post-test) and the Student t-test is that the latter uses the raw quality values returned by the target algorithm, while the former transforms the values into ranks. Hence, the F-test can detect minimal but consistent differences between the performance of the configurations but it is insensitive to large sporadic differences, whereas the t-test is influenced by such outliers. Figure 5 shows box-plots comparing the configurations obtained using both statistical tests and Table 5 provides the Wilcoxon test p-values. The first set of plots show the average performance of the candidates on the test set and the second set of plots compares the average performance of the candidates per instance. The results of the Wilcoxon test indicate significant differences only for the MOACO case, where the usage of the t-test leads to better performance. It is interesting, however, to analyze in more detail the SPEAR configuration scenario. While there is no significant difference w.r.t. to the average performance (mean runtime), the F-test leads to shorter runtimes on more instances than the t-test; however, the t-test performs much better than the configurations obtained by the F-test on the subset of the *hsat* instances. Configurations obtained by using the F-test solve a majority of instances faster than configurations obtained by using the t-test; this difference is statistically significant. This corresponds to the fact that the F-test prefers to improve the mean ranking by performing well on a majority of instances while the t-test improves the mean performance and tends to reduce worst case performance, which in the SPEAR configuration scenario are very long runtimes. In this sense, these results confirm earlier observations for different configurators [11, 16].

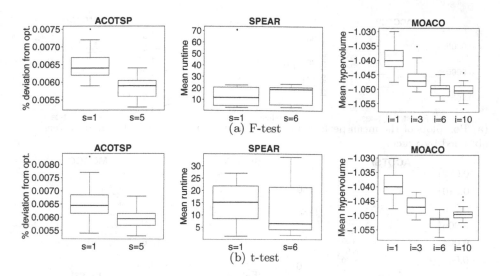

(a) F-test

(b) t-test

Fig. 4. Box plots of the mean performace over the test instances of 20 configurations obtained by `irace` using $N^{max} \in \{1, \text{default}\}$

Table 4. Wilcoxon signed-rank test p-values comparing configurations obtained by `irace` using the default setting of N^{max} vs. $N^{max} = 1$, over the test set

	ACOTSP	SPEAR	MOACO
F-test	$3.624e^{-5}$	0.7285	0.4304
t-test	0.0005856	0.5958	0.4304

Table 5. Wilcoxon signed-rank test p-values comparing configurations obtained by `irace` using F-test vs. t-test

ACOTSP	SPEAR	MOACO
0.2943	0.5958	0.03277

Statistical Test Confidence Level. The confidence level of the `irace` elimination test is set by default to 0.95. Larger values mean that the test is more strict, so it takes more evaluations (or clearer differences) to eliminate configurations; lower values allow eliminating configurations faster, save budget, but risk removing good configurations based on few unlucky runs. We assess the effect of this parameter on the configuration process by experimenting with confidence levels $\in \{0.75, 0.95, 0.99\}$. Results are summarized in Fig. 6 and Table 6. For the ACOTSP configuration scenario, a setting of 0.99 is clearly worse than the default setting. Even if on the MOACO configuration scenario the 0.99 confidence level is significantly better than the default, the absolute difference is small and we would recommend using the default 0.95 level. Considering a smaller confidence level such as 0.75 may be an option. In fact, in two cases this setting is

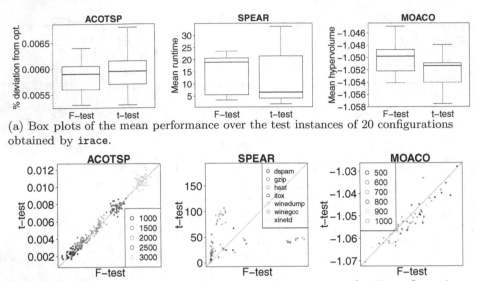

(a) Box plots of the mean performance over the test instances of 20 configurations obtained by irace.

(b) Scatter plots of the mean performace per instance over the 20 configurations obtained by irace.

Fig. 5. Comparison of the mean performance over the test instances of 20 configurations obtained by irace using F-test and t-test

statistically better than the default setting while in one it is worse. However, the results also indicate that the behavior of irace is affected differently by the confidence level used depending on the statistical test used (see, e.g. MOACO configuration scenario). This is different from the other experiments, where the impact of irace parameter settings was similar for both elimination tests.

5   Final Remarks and Future Work

In this paper, we analyse the impact of five irace parameters on the final configuration performance. The experiments were performed on three configuration scenarios. The ACOTSP and the MOACO scenarios show a fairly homogeneous training set, while the SPEAR configuration scenario has a highly heterogeneous set of instances and a large variability of the quality values (runtime), making the configuration process more variable. The default settings of the number of iterations and the number of elite configurations proved to be reasonably robust. Reducing the setting of the first elimination test did not improve the performance of irace, although the results obtained suggest that lower values for this parameter could be used with highly homogeneous sets of instances. Larger differences were observed when altering the type of statistical test. However, in this case, the best setting depends on the goal of the configuration process. If the goal should take into account outliers, then irace should use the t-test rather than the F-test. Finally, the confidence level had a strong effect on the results.

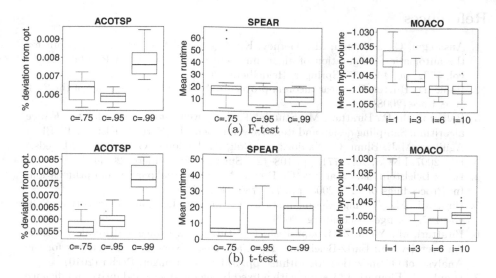

Fig. 6. Box plots of the mean performace over the test instances of 20 configurations obtained by `irace` using confidence level in $\{0.75, 0.95, 0.99\}$

Table 6. Wilcoxon signed-rank test p-values comparing configurations obtained by `irace` using confidence level in $\{0.75, 0.95, 0.99\}$

	0.75 vs. 0.95			0.99 vs. 0.95		
	ACOTSP	SPEAR	MOACO	ACOTSP	SPEAR	MOACO
F-test	0.01531	0.4091	$1.907e^{-6}$	$1.907e^{-6}$	0.7841	0.002325
t-test	0.02148	0.9563	0.1429	$1.907e^{-6}$	0.3683	0.2455

Large values were consistently worse, whereas lower values were sometimes better depending on the scenario and the type of statistical test. Further work will extend and complement the current experimental analysis in order to account for more parameter settings and their possible interactions. Additional scenarios may help to identify clearer trends or use an automatic configuration process to configure improved default settings of `irace`. Nonetheless, the insights obtained in this work are helping us to design future improvements to `irace`.

Acknowledgments. This work received support from the META-X project, an *Action de Recherche Concertée* funded by the Scientific Research Directorate of the French Community of Belgium, the COMEX project within the Interuniversity Attraction Poles Programme of the Belgian Science Policy Office, and the EU FP7 ICT Project COLOMBO, Cooperative Self-Organizing System for Low Carbon Mobility at Low Penetration Rates (agreement no. 318622). Manuel López-Ibáñez and Thomas Stützle acknowledge support from the Belgian F.R.S.-FNRS, of which they are a postdoctoral researcher and a senior research associate, respectively.

48   L. Pérez Cáceres, M. López-Ibáñez, and T. Stützle

References

1. Ansótegui, C., Sellmann, M., Tierney, K.: A gender-based genetic algorithm for the automatic configuration of algorithms. In: Gent, I.P. (ed.) CP 2009. LNCS, vol. 5732, pp. 142–157. Springer, Heidelberg (2009)
2. Babić, D., Hutter, F.: Spear theorem prover. In: SAT 2008: Proceedings of the SAT 2008 Race (2008)
3. Balaprakash, P., Birattari, M., Stützle, T.: Improvement strategies for the F-race algorithm: Sampling design and iterative refinement. In: Bartz-Beielstein, T., Blesa Aguilera, M.J., Blum, C., Naujoks, B., Roli, A., Rudolph, G., Sampels, M. (eds.) HM 2007. LNCS, vol. 4771, pp. 108–122. Springer, Heidelberg (2007)
4. Bartz-Beielstein, T., Lasarczyk, C., Preuss, M.: Sequential parameter optimization. In: Proceedings of CEC 2005, pp. 773–780. IEEE Press (2005)
5. Birattari, M.: Tuning Metaheuristics: A Machine Learning Perspective. SCI, vol. 197. Springer, Heidelberg (2009)
6. Birattari, M., Yuan, Z., Balaprakash, P., Stützle, T.: F-race and iterated F-race: An overview. In: Bartz-Beielstein, T., et al. (eds.) Experimental Methods for the Analysis of Optimization Algorithms, pp. 311–336. Springer, Berlin (2010)
7. Branke, J., Elomari, J.: Racing with a fixed budget and a self-adaptive significance level. In: Nicosia, G., Pardalos, P. (eds.) LION 7. LNCS, vol. 7997, pp. 272–280. Springer, Heidelberg (2013)
8. Conover, W.J.: Practical Nonparametric Statistics. John Wiley & Sons (1999)
9. Hoos, H.H.: Automated algorithm configuration and parameter tuning. In: Hamadi, Y., et al. (eds.) Autonomous Search, pp. 37–71. Springer, Berlin (2012)
10. Hutter, F., Hoos, H.H., Leyton-Brown, K.: Sequential model-based optimization for general algorithm configuration. In: Coello Coello, C.A. (ed.) LION 5. LNCS, vol. 6683, pp. 507–523. Springer, Heidelberg (2011)
11. Hutter, F., Hoos, H.H., Leyton-Brown, K., Stützle, T.: ParamILS: An automatic algorithm configuration framework. Journal of Artificial Intelligence Research 36, 267–306 (2009)
12. López-Ibáñez, M., Dubois-Lacoste, J., Stützle, T., Birattari, M.: The irace package, iterated race for automatic algorithm configuration. Tech. Rep. TR/IRIDIA/2011-004, IRIDIA, Université Libre de Bruxelles, Belgium (2011)
13. López-Ibáñez, M., Stützle, T.: The automatic design of multi-objective ant colony optimization algorithms. IEEE Transactions on Evolutionary Computation 16(6), 861–875 (2012)
14. Schneider, M., Hoos, H.H.: Quantifying homogeneity of instance sets for algorithm configuration. In: Hamadi, Y., Schoenauer, M. (eds.) LION 6. LNCS, vol. 7219, pp. 190–204. Springer, Heidelberg (2012)
15. Siegel, S., Castellan Jr., N.J.: Non Parametric Statistics for the Behavioral Sciences, 2nd edn. McGraw Hill (1988)
16. Smit, S.K., Eiben, A.E.: Beating the "world champion" evolutionary algorithm via REVAC tuning. In: Ishibuchi, H., et al. (eds.) Proceedings of CEC 2010, pp. 1–8. IEEE Press (2010)
17. Stützle, T.: ACOTSP: A software package of various ant colony optimization algorithms applied to the symmetric traveling salesman problem (2002), http://www.aco-metaheuristic.org/aco-code/
18. Zitzler, E., Thiele, L., Laumanns, M., Fonseca, C.M., Grunert da Fonseca, V.: Performance assessment of multiobjective optimizers: an analysis and review. IEEE Transactions on Evolutionary Computation 7(2), 117–132 (2003)

An Improved Multi-objective Algorithm for the Urban Transit Routing Problem

Matthew P. John[1,2], Christine L. Mumford[1], and Rhyd Lewis[2]

[1] Cardiff School of Computer Science & Informatics, UK
[2] Cardiff School of Mathematics, UK
JohnMP@cardiff.ac.uk

Abstract. The determination of efficient routes and schedules in public transport systems is complex due to the vast search space and multiple constraints involved. In this paper we focus on the Urban Transit Routing Problem concerned with the physical network design of public transport systems. Historically, route planners have used their local knowledge coupled with simple guidelines to produce network designs. Several major studies have identified the need for automated tools to aid in the design and evaluation of public transport networks. We propose a new construction heuristic used to seed a multi-objective evolutionary algorithm. Several problem specific mutation operators are then combined with an NSGAII framework leading to improvements upon previously published results.

1 Introduction

The Urban Transit Network Design Problem (UTNDP) involves the determination of an efficient set of routes and schedules for public transportation systems such as bus, rail and tram networks. Ceder and Wilson [1] identified five main stages for bus service planning: network design, frequency setting, timetable development, bus scheduling and driver scheduling. Given that each stage of the UTNDP is NP-hard [2], it is usually considered impractical to solve all the stages simultaneously.

In this paper we focus on the network design element, which is tasked with determining an efficient set of routes on an already established road (or rail) network, usually with previously identified pickup and drop off locations (e.g. bus stops). Building upon recently published work by Mumford [3], we present a much improved multi-objective approach based upon NSGAII [4]. Our approach considers the trade offs between passenger and operator costs by producing approximate Pareto optimal sets for consideration by a human decision maker. Furthermore, we provide a discussion of the many specialised heuristics and operators we have tested during the development of our approach.

Historically, route planners have used a combination of local knowledge and simple guidelines to produce route sets. Several major studies (see [5, 6]) have identified the need for automated computer based tools for the design and evaluation of public transport networks. Automation is, however, highly complex

C. Blum and G. Ochoa (Eds.): EvoCOP 2014, LNCS 8600, pp. 49–60, 2014.
© Springer-Verlag Berlin Heidelberg 2014

and computationally expensive due to the large search space and multiple constraints involved in public transportation planning. The increase in congestion, pollution, greenhouse gas emissions and dwindling oil resources have placed emphasis on the use of public transport in recent years in an attempt to reduce the reliance of the private car. Achieving an increase in public transportation usage is clearly desirable but is also an extremely complex issue. However frequent and reliable cost-effective services are clearly key attributes.

Bagloee and Ceder [7] have recently pointed out that many public transit networks have not been reappraised from anywhere between 20 to 50 years. Land use patterns have changed considerably in this time period with the migration away from town centres into surrounding suburban areas; however public transport has been relatively slow to respond. It is our view that the development of automated tools to aid public transport networks is timely.

Prior to 1979, the few papers published on the UTNDP considered only highly specific problem instances [8, 9]. In 1979, Christoph Mandl [10, 11] approached the problem in a rather more generic form. He concentrated on the network design phase, and developed a two-stage solution. First a set of feasible routes is generated, second, heuristics are applied to improve the quality of the routes in this set. Following Mandl's pioneering work, heuristic methods have been widely used to solve the UTNDP, e.g. [1, 12]. With the advancement of computing technology over the last two decades however, metaheuristic techniques have become increasingly popular for solving these problems, particularly genetic algorithms (GAs) [13–16]. Other metaheuristic methods such as tabu search and simulated annealing can also be seen in [17, 18]. Nevertheless, comparative work has been limited to Mandl's 15 vertex instance.

One of the first approaches using a GA was proposed by Pattnaik et al. [15] utilising a route set generation procedure guided by a demand matrix, designer knowledge, and route constraints to produce a set of candidate routes. The role of the GA is simply to select a number of routes from the candidate set, where each route is given a unique identifier encoded in binary. This approach is similar to those proposed by Tom and Mohan [16] and Fan and Machemehl [19].

Chakroborty and Dwivedi [14] generated an initial set of candidate route sets each with a fixed number of routes. A GA was then applied using two crossover operators: 1) Inter-string crossover which exchanges routes from the parent route sets, and 2) Intra-string crossover which exchanges parts of a route in a parent if two routes share a common vertex. Mutation is then applied by randomly selecting a vertex and changing it to any of its acceptable adjacent vertices.

Similar to Chakroborty and Dwivedi, Fan et al. [20] generated an initial population of feasible route sets. Their GA used a mutation operator where a vertex can be either added to the end of a route or removed from the start. However, their GA lacked a crossover operator.

Bagloee and Ceder [7] tackled real sized road networks using a combination of heuristics, a GA and an ant-system. They determined the location of stops based upon the distance to high concentrations of travel demand and then used a system inspired by Newtonian gravity theory to produce a set of candidate routes. A GA was then used to search through the candidate routes to find a good solution – the frequencies of the routes was computed simultaneously.

Recently Mumford [3] presented an approach using a constructive heuristic to generate an initial population. This was then evolved using a multi-objective evolutionary algorithm (MOEA). Mumford's crossover operator alternates the selection of routes between two parents, favoring routes containing vertices that are not currently present in the partial solution, while ensuring route connectivity. Two mutation operators are also proposed that add or delete a bounded random number of vertices from a route set. Four benchmark instances based on real-world bus route networks were also created and made publicly available [3].

2   Problem Description

The network design problem can be formally stated as follows. Given a graph $G = (V, E)$ where $V = \{v_1, \ldots, v_n\}$ is a set of vertices and $E = \{e_1, \ldots, e_m\}$ is a set of edges, we are given:

- A weight for each edge, W_{e_i}, which defines the time it takes to traverse edge e_i;
- A matrix $\mathbf{D}_{n \times n}$ where D_{v_i, v_j} gives the passenger demand between a pair of vertices v_i and v_j.

A route R_i is defined as a simple path (i.e. no loops/repeated vertices) through the graph G. Let $G_{R_i} = (V_{R_i}, E_{R_i})$ be the subgraph induced by a route R_i. A solution is defined as a set of overlapping routes $\mathcal{R} = \{R_1, \ldots, R_r\}$ where the number of routes, r, and the minimum, m_1, and maximum, m_2, number of vertices in a route are to be specified by the user. In order for $\mathcal{R}$ to be valid the following conditions must hold:

$$\bigcup_{i=1}^{|\mathcal{R}|} V_{R_i} = V \tag{1}$$

$$m_1 \leq |V_{R_i}| \leq m_2 \ \forall R_i \in \mathcal{R} \tag{2}$$

$$G_{\mathcal{R}} = \left(\bigcup_{i=1}^{|\mathcal{R}|} V_{R_i}, \bigcup_{i=1}^{|\mathcal{R}|} E_{R_i}\right) \text{ is connected} \tag{3}$$

$$|\mathcal{R}| = r \tag{4}$$

Constraint (1) ensures that all vertices in V are in at least one route in $\mathcal{R}$. Constraint (2) specifies that each route should contain between m_1 and m_2 vertices (these values are based on considerations such as driver fatigue and the difficulty of maintaining the schedule [6]). Constraint (3) specifies that a path exists between all pairs of vertices in $G_{\mathcal{R}}$. If Constraint (1) is satisfied then

$G_{\mathcal{R}} = (V, \bigcup_{i=1}^{|\mathcal{R}|} E_{R_i})$. Finally, Constraint (4) ensures that the solution contains the correct number of routes.

For this problem formulation, the following assumptions are also made:

1. There will always be sufficient vehicles on each route $R_i \in \mathcal{R}$ to ensure that the demand between every pair of vertices is satisfied.
2. A vehicle will travel back and forth along the same route, reversing its direction each time it reaches a terminal vertex.
3. The transfer penalty (representing the inconvenience of moving from one vehicle to another) is set at a fixed constant. In this study a value of 5 minutes is used in line with previous studies (e.g. [14, 21]).
4. Passenger choice of routes is based on shortest travel time (which includes transfer penalties).

In this problem we consider both the *passenger cost* and *operator cost*. In general, passengers would like to travel to their destination in the shortest possible time, but avoiding the inconvenience of making too many transfers. We define a shortest path between two vertices using the route set $\mathcal{R}$ as $\alpha_{v_i,v_j}(\mathcal{R})$. A path may include both transport links and transfer links (a transfer link facilitates the changing from one vehicle to another with the associated time penalty). This is shown in Fig. 1 with the original network expanded to include transfer vertices and transfer links. The shortest path evaluation is thus completed on the transit network Fig. 1(b). The minimum journey time, $\alpha_{v_i,v_j}(\mathcal{R})$, from any given pair of vertices is thus made up of two components: in vehicle travel time and transfer penalty. We define the *passenger cost* for a route set $\mathcal{R}$ to be the mean journey time over all passengers as given by Mumford [3]:

$$F_1(\mathcal{R}) = \frac{\sum_{i,j=1}^{n} D_{v_i,v_j} \alpha_{v_i,v_j}(\mathcal{R})}{\sum_{i,j=1}^{n} D_{v_i,v_j}} \tag{5}$$

Operator costs depend on many factors, such as the number of vehicles needed to maintain the required level of service, the daily distance travelled by the vehicles and the costs of employing sufficient drivers. We use a simple proxy for operator costs: the sum of the costs (in time) for traversing all the routes in one direction, as given by Mumford [3]:

$$F_2(\mathcal{R}) = \sum_{\forall R_i \in \mathcal{R}} \sum_{\forall e_j \in R_i} W_{e_j} \tag{6}$$

3 Methodology

In this paper we propose an approach that seeds a MOEA with a high quality initial population formed using a powerful heuristic construction procedure. NSGAII [4] is then used with the crossover and repair operators proposed by Mumford [3] along with several new mutation operators.

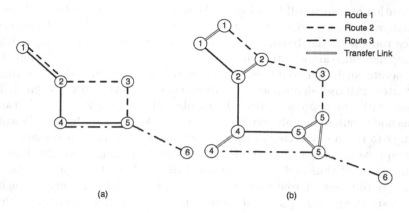

Fig. 1. (a) Route network – road network with routes overlayed (b) Transit network – network used for evaluation

3.1 NSGAII

NSGAII is an elitist non-dominated sorting MOEA widely used to solve multi-objective optimization problems. It has been shown to find a better spread of solutions and convergence nearer to the true Pareto-optimal front compared with other Pareto based methods [4] .

The basic form of an NSGAII generation proceeds by creating an offspring population of size N. This is then combined with the parent population of size N to produce a population, $P = \{\mathcal{R}_1, \mathcal{R}_2, \dots, \mathcal{R}_{2N}\}$. Let us now define two attributes of a route set $\mathcal{R}_i$: 1) $\mathcal{R}_{i_{\text{rank}}}$ the non-dominated front that $\mathcal{R}_i$ belongs to, and 2) $\mathcal{R}_{i_{\text{dist}}}$ the crowding distance associated with $\mathcal{R}_i$ as defined by Deb et al. [4]. P is then sorted such that $\forall \mathcal{R}_i, \mathcal{R}_j \in P$ $\mathcal{R}_{i_{\text{rank}}} \leq \mathcal{R}_{j_{\text{rank}}}$ and $\mathcal{R}_{i_{\text{dist}}} \geq \mathcal{R}_{j_{\text{dist}}}$ for $i < j$. The successor population is then formed by taking the first N solutions in P.

Similar to [4], in our case a new population is generated using binary tournament selection with a crossover probability of 0.9. The probability of mutating each route in an offspring is set as $\frac{1}{r}$.

3.2 Heuristic Construction

As mentioned earlier, heuristic construction was widely used for the network design stage before the use of metaheuristic algorithms became more widespread in the literature. Our construction heuristic creates solutions by incorporating knowledge that exploits the underlying structure of the problem.

The majority of previous approaches for tackling network design (see [18, 19, 15]) generate a candidate pool of routes using Dijkstra's shortest path algorithm and Yen's [22] k-shortest path algorithm to enumerate all possible routes. An optimization algorithm is then applied to find the best combination of routes from the candidate pool, although there is still no guarantee that the optimal configuration of routes will be selected. As such we have chosen to combine our approach with that used by Shih and Mahmassani [21] with modifications to produce route sets that balance the cost to the operator and passengers whilst ensuring the constraints of the problem are adhered to. Shih and Mahmassani [21] generate an initial route set using a heuristic procedure that continually adds routes until user-defined levels on directness and coverage are reached. This contrasts with our approach where the number of routes is fixed. Furthermore, our method ensures that all vertices are present within a route set, allowing a passenger to reach any vertex in the network using transfers if necessary.

Our approach generates a set of weighted graphs using a weighted sum of normalized travel time and (1 − normalized demand) with weights specified in advance by the user. A route set, $\mathcal{R}$, is then generated from a given weighted graph. A spanning subgraph is first created using an iterative procedure with the objective of minimizing the sum of the weighted edge costs. In the first iteration of this procedure the pair of vertices (seed pair) with the lowest weighted edge cost are selected. In subsequent iterations a seed pair contains one vertex already contained in the subgraph, ensuring that routes remain connected. A seed pair is then expanded to form a route by adding adjacent vertices.

During this expansion process vertices that are adjacent to the first or last vertex in the route are sorted against their weighted edge cost. The minimum cost unused vertex is then added to the route. If there are no unused vertices the minimum cost adjacent vertex is instead added to the route providing of course that the vertex is not present in the current route under construction. If multiple vertices are in the set of potential vertices with equal weight a random vertex from this set is selected. Vertices are continually inserted until a vertex can no longer be inserted that would not cause a constraint to be violated.

Providing that $|\mathcal{R}| < r$ the next stage of the heuristic procedure is applied using an approach utilized by Shih and Mahmassani [21]. The vertex pairs (v_i, v_j) that are yet to be satisfied directly (i.e. it is not possible to travel between the vertices without having to make a transfer) are extracted from the network and sorted in non-ascending order based upon the demand D_{v_i, v_j}. The unsatisfied vertex pairs are taken in order and Yen's k-shortest path algorithm is applied to determine if a valid route, originating at v_i and terminating at v_j, can be constructed that obeys all of the constraints. In our case a maximum of ten 'shortest' paths are explored for each vertex pair. An alternative to satisfying the vertex pairs that are not satisfied directly is to minimize the travel time for passengers on high demand vertex pairs. In this instance all the vertex pairs are extracted and sorted based upon non-ascending demand. If a valid route is found, the cost of the route is calculated and compared with the $\alpha_{v_i, v_j}(\mathcal{R})$. If the cost is less than $\alpha_{v_i, v_j}(\mathcal{R})$ the route is inserted into $\mathcal{R}$. This process is applied iteratively until $|\mathcal{R}| = r$.

3.3   Genetic Operators

Crossover: We use the crossover operator proposed by Mumford [3], which ensures that the problem constraints are obeyed. Given two parents, the crossover operator constructs an offspring, $\mathcal{R}'$, by alternatively selecting a route from each parent that maximises the proportion of unseen vertices, until $|\mathcal{R}'| = r$. The set of unseen vertices is defined as $V_{\text{unseen}} = V - V_{\mathcal{R}'}$. Consider a route $R_i = \langle 1, 7, 8, 9, 12, 14 \rangle$ that is contained in one of the parents and is being considered for insertion into $\mathcal{R}'$. If $\mathcal{R}' = \{R_1\}$ where $R_1 = \langle 12, 15, 0, 5, 3 \rangle$ then $V_{\text{unseen}} = R_i - R_1 = \{1, 7, 8, 9, 14\}$. Therefore the proportion of unseen vertices is $\frac{|V_{\text{unseen}}|}{|R_i|} = \frac{5}{6}$ in this case.

After crossover has been applied it is possible that the offspring will not contain all the vertices in V. In these cases the repair procedure used by Mumford is applied that attempts to add the missing vertices to either the back or front of the routes.

Mutation: In our approach eight mutation operators are used. Some of these apply heuristics to mutate the route set in a way that encourages an improvement in quality. Mutation must be carefully controlled to prevent violation of the problem constraints. The names of these mutation operators are *add-nodes*, *del-nodes*, *exchange*, *merge*, *replace*, *remove-overlapping*, *two-opt* and *invert-exchange*.

Add-nodes and *del-nodes* were both proposed in [3]. At the start of *add-nodes* or *del-nodes* an integer I is generated uniform randomly in the range $[1, r \times \frac{m_2}{2}]$, giving the number of vertices to be added or removed from $\mathcal{R}$. A route $R_i \in \mathcal{R}$ is then selected at random and, in the case of *add-nodes*, vertices are added to the end of the route until the addition of a vertex would cause Constraint (2) to be violated or, result in R_i no longer being a simple path. Following a similar approach, vertices are then added to the front of the route if possible. This process is repeated for each $R_i \in \mathcal{R}$ until I vertices have been added to $\mathcal{R}$ or all routes have been exhausted. The case is the same for *del-nodes* with I vertices being removed from $\mathcal{R}$ whilst ensuring feasibility.

The *Exchange* operator, as proposed by Mandl [10], selects a route at random. The route set is then searched to determine if there exists a route with a common vertex to the selected route. The problem constraints are also checked to determine if $\mathcal{R}$ will be valid after the mutation has been applied. If valid, the two routes are split at the first common vertex, creating four route segments. The two original routes are then replaced by exchanging the segments to create two new routes. The exchange of route parts attempts to reduce the number of transfers passengers must make.

Similar to *exchange*, the *merge* operator, selects a random route and searches the remaining routes to find a route that shares a common terminal vertex. The two routes are then merged creating one continuous route, disregarding one of the common terminal vertices – providing that Constraint (2) is not violated and the merged route is a simple path. If successful, a route generation procedure (*route-gen*) is then used to generate a new route for insertion. The route generation procedure is as follows: vertex pairs that are not yet satisfied directly are extracted from the route set and sorted via non-ascending demand (i.e. given every pair of vertices in the network we are only interested in those that cannot be reached without transfers, given the current configuration of $\mathcal{R}$).

Yen's k-shortest path algorithm is then used to generate a bounded number of paths, in our case ten, between the two vertices and a random path selected for insertion. If the number of vertex pairs yet to be satisfied directly is zero, a path is generated between the vertices with the highest demand in the network.

The *Replace* operator removes a route $R_i \in \mathcal{R}$ that satisfies the least demand directly compared with all other routes in $\mathcal{R}$. A replacement route is then generated using the *route-gen* procedure described above. The purpose of the *replace* mutation is to sacrifice routes that serve a relatively low demand in place of high demand routes. *Replace* can cause a route set to become invalid if the removed route acted as a transfer hub for routes, i.e. the route set was only connected when the removed route was present. If this situation occurs the repair procedure used during crossover is applied and, if successful, the mutated solution is returned. Otherwise the mutation is abandoned.

The *Remove-overlapping* operator replaces a route that is a subset of another route. If an overlapping route is discovered, it is removed and the route generation procedure described above is used to produce a replacement route. Replacing the route provides the operator with the ability to remove duplicate services and use these resources to serve other passenger demand.

Two-opt, proposed in 1958 by Croes [23] for use with the traveling salesman problem selects two vertices at random in a route and inverts the vertices between them. Its original purpose was to remove crossover points in a route, however this is not allowed to occur in this context. In our case it reorders the vertices in a route attempting to reduce the travel time between vertex pairs, abandoning infeasible attempts.

Invert-exchange selects two routes at random and generates two random index locations. The vertices between the two random index locations are then inverted and exchanged between the two routes. For example, given two routes $R_1 = \langle 3, 5, 8, 10, 12, 15 \rangle$ and $R_2 = \langle 1, 6, 9, 8, 11, 7 \rangle$ with the selected indices of 3 and 5. We invert everything in R_1 between the indices giving $\langle 12, 10, 8 \rangle$ then replace the vertices in R_2 between the indices with the inverted section from R_1. In this case the resultant two routes would be $R_3 = \langle 3, 5, 11, 8, 9, 15 \rangle$ and $R_4 = \langle 1, 6, 12, 10, 8, 7 \rangle$. *Invert-exchange* attempts to decrease the travel time between vertices and prevent passengers having to make a transfer. Similar to *two-opt* there is a high possibility that the majority of routes created using this approach will be infeasible. As such, two routes are continually chosen at random until the routes have been exhausted or a feasible solution has been found.

4   Results

In this section we show how the algorithm of Mumford [3] (Algorithm A), based on the SEAMO2 framework, can be improved, by seeding the MOEA with our heuristically generated solutions (Algorithm B). We then look at the effects of adding our mutation operators (Algorithm C) and finally look at the effects of using the NSGAII framework (Algorithm D) as opposed to SEAMO2. All experiments use an initial population of size 200 and are run for 200 generations, the

Table 1. Problem instances used for comparison with the lower bound (LB) for each objective

Instance	Vertices	Edges	r	m_1	m_2	Vertices in Typical Transit Net.	LB_{F_1}	LB_{F_2}
Mandl	15	20	6	2	8	$6 \times \frac{(2+8)}{2} = 30$	10.0058	63
Mumford0	30	90	12	2	15	102	13.0121	94
Mumford1	70	210	15	10	30	300	19.2695	294
Mumford2	110	385	56	10	22	896	22.1689	749
Mumford3	127	425	60	12	25	1110	24.7453	928

same as in [3] so that valid comparisons can be made. Running times range from a couple of seconds for Mandl's instance up to two days for Mumford3. Twenty replicate runs are used and the results have been combined into approximate Pareto sets for comparison. Problem instances generated by Mumford [3] along with Mandl's [10] benchmark are used. Table 1 summarises the details of each instance along with the parameters used and lower bounds as given in [3].

Using our heuristic construction procedure, a subset of unique solutions are randomly selected for insertion into an initial population. Randomly generated solutions are then used to top-up the initial population if there are too few heuristic solutions. These random solutions are created using the same approach as Mumford [3] to seed her MOEA. In summary this approach constructs a route set one route at a time. A route length is randomly generated between m_1 and m_2 and a random vertex, $v \in V$, is selected as the seed. A randomly selected adjacent vertex is then added to the back of the route, and this process is repeated. Once the vertices that can be added to the back are exhausted the process is repeated from the front of the route until the desired length is achieved.

We firstly augment Mumford's SEAMO2 algorithm with our heuristic method for generating the initial population (Algorithm B). Table 2 presents the best

Table 2. Best objective values extracted from twenty replicate runs using heuristic seeding for the initial population (Algorithm B). Mumford's [3] results are given in brackets.

		Mandl	Mumford0	Mumford1	Mumford2	Mumford3
Best for passenger	F_1	**10.25**(10.33)	**15.40**(16.05)	**23.91**(24.79)	**27.02**(28.65)	**29.50**(31.44)
	F_2	212(224)	745(759)	1861(2038)	5461(5632)	6320(6665)
Best for operator	F_1	13.48(15.13)	32.78(32.40)	39.98(34.69)	32.33(36.54)	36.12(36.92)
	F_2	**63**(63)	**95**(111)	**462**(568)	**1875**(2244)	**2301**(2830)

Table 3. S-metric comparison over the five benchmark instances for our proposed modifications

Instance	Alg. A	Alg. B	Alg. C	Alg. D
Mandl	2620.19	2620.21	2626.94	2631.16
Mumford0	14951.24	15031.81	15304.90	15451.50
Mumford1	111947.82	114614.99	114972.74	117866.22
Mumford2	306261.94	322753.72	322618.47	337987.85
Mumford3	507983.61	538371.55	539296.98	562793.74

solutions from the passenger and operator perspective compared to the findings of [3]. We see that our heuristic is clearly beneficial, producing an improvement over all the instances. Where an improvement is made in the objective value, either from the passenger or operator perspective, we can see that an improvement is also made to the other objective on the majority of instances. An improvement from the operator perspective on Mandl's instance is not possible as 63 is the lower bound for this instance [3]. However, a decrease is observed in the

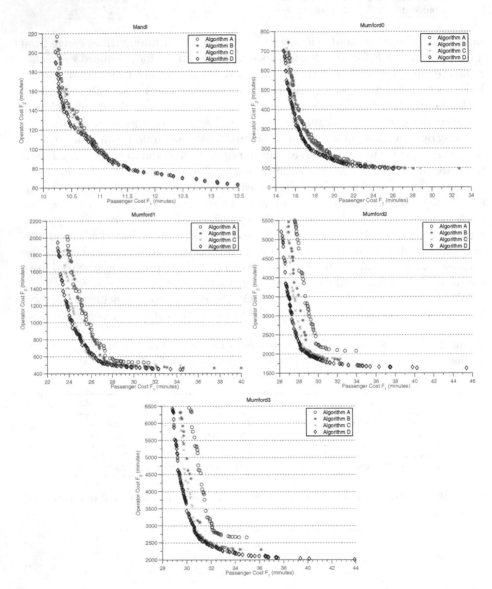

Fig. 2. Combined Pareto fronts extracted from twenty runs for each of the five benchmark instances using the four algorithms

passenger objective associated with the operator objective of 63 compared to that found by Mumford.

Taking the best algorithm (Algorithm B) from the previous experiment we now examine the effect of augmenting the algorithm with our proposed mutation operators. Comparing S-metric values for Algorithms B and C (Table 3) we can see that an improvement is achieved with all instances apart from Mumford2. If the Pareto set is plotted for Mumford2 (Fig. 2) it can be seen that there is an improvement in the passenger objective for the majority of solutions. However, we struggle to make improvements in the extremes of the operator objective.

Given the popularity of NSGAII and its stated ability to produce a Pareto set closer to the true Pareto-optimal front compared with other Pareto based methods [4], it was used instead of SEAMO2 in our third set of experiments. Here, we used our mutation operators and heuristic seeding, together with Mumford's crossover operator. As mentioned earlier a probability of crossover and mutation of 0.9 and $\frac{1}{r}$ respectively are used. A comparison of S-metric values, Table 3, shows that NSGAII gives an improvement over all the problem instances. This is displayed graphically in Fig. 2. These improvements can be attributed to the following: 1) A higher selection pressure compared with SEAMO2, and 2) Increased rate of mutation leading to a greater exploration of the search space.

5   Conclusion

This paper has presented a new construction heuristic for creating initial solutions to the transit network design problem. For this problem it has been shown that the use of heuristic solutions in a MOEA's initial population can produce an improvement in the resultant Pareto set – compared with the use of solely randomly generated solutions. Several mutation operators have also been proposed and combined with our construction heuristic to produce an improvement over previously published results. Finally, we have also shown that further improvements can be found if NSGAII is used in place of SEAMO2. We are currently investigating the introduction of frequency setting to the MOEA framework using more realistic evaluation, allowing our model to more accurately reflect passenger choice. In addition we are collecting real-world data to enable the production of more benchmark instances.

References

1. Ceder, A., Wilson, N.H.M.: Bus network design. Transportation Research Part B 20(4), 331–344 (1986)
2. Magnanti, T.L., Wong, R.T.: Network design and transportation planning: Models and algorithms. Transportation Science 18(1), 1–55 (1984)
3. Mumford, C.L.: New heuristic and evolutionary operators for the multi-objective urban transit routing problem. In: 2013 IEEE Congress on Evolutionary Computation (CEC), pp. 939–946 (2013)

4. Deb, K., Pratap, A., Agarwal, S., Meyarivan, T.: A fast and elitist multiobjective genetic algorithm: NSGA-II. IEEE Transactions on Evolutionary Computation 6(2), 182–197 (2002)
5. Nielsen, G., Nelson, J.D., Mulley, C., Tegner, G., Lind, G., Lange, T.: Public transport–planning the networks. HiTrans Best Practice Guide (2005)
6. Zhao, F., Gan, A.: Optimization of transit network to minimize transfers (2003)
7. Bagloee, S.A., Ceder, A.A.: Transit-network design methodology for actual-size road networks. Transportation Research Part B 45(10), 1787–1804 (2011)
8. Lampkin, W., Saalmans, P.D.: The design of routes, service frequencies, and schedules for a municipal bus undertaking: A case study. In: OR, pp. 375–397 (1967)
9. Silman, L.A., Barzily, Z., Passy, U.: Planning the route system for urban buses. Computers & Operations Research 1(2), 201–211 (1974)
10. Mandl, C.E.: Applied network optimization. Academic Pr. (1979)
11. Mandl, C.E.: Evaluation and optimization of urban public transportation networks. European Journal of Operational Research 5(6), 396–404 (1980)
12. Baaj, M.H., Mahmassani, H.S.: Hybrid route generation heuristic algorithm for the design of transit networks. Transportation Research Part C 3(1), 31–50 (1995)
13. Agrawal, J., Mathew, T.V.: Transit route network design using parallel genetic algorithm. Journal of Computing in Civil Engineering 18(3), 248–256 (2004)
14. Chakroborty, P., Dwivedi, T.: Optimal route network design for transit systems using genetic algorithms. Engineering Optimization 34(1), 83–100 (2002)
15. Pattnaik, S.B., Mohan, S., Tom, V.M.: Urban bus transit route network design using genetic algorithm. Journal of Transportation Engineering 124(4), 368–375 (1998)
16. Tom, V.M., Mohan, S.: Transit route network design using frequency coded genetic algorithm. Journal of Transportation Engineering 129(2), 186–195 (2003)
17. Fan, W., Machemehl, R.B.: A tabu search based heuristic method for the transit route network design problem. Computer-aided Systems in Public Transport, 387–408 (2008)
18. Fan, W., Machemehl, R.B.: Using a simulated annealing algorithm to solve the transit route network design problem. Journal of Transportation Engineering 132(2), 122–132 (2006)
19. Fan, W., Machemehl, R.B.: Optimal transit route network design problem with variable transit demand: genetic algorithm approach. Journal of Transportation Engineering 132(1), 40–51 (2006)
20. Fan, L., Mumford, C.L., Evans, D.: A simple multi-objective optimization algorithm for the urban transit routing problem. In: IEEE Congress on Evolutionary Computation, CEC 2009, pp. 1–7 (2009)
21. Shih, M.C., Mahmassani, H.S.: A design methodology for bus transit networks with coordinated operations. Technical Report SWUTC/94/60016-1 (1994)
22. Yen, J.Y.: Finding the k shortest loopless paths in a network. Management Science 17(11), 712–716 (1971)
23. Croes, G.A.: A method for solving traveling-salesman problems. Operations Research 6(6), 791–812 (1958)

An Iterated Greedy Heuristic
for Simultaneous Lot-Sizing and Scheduling
Problem in Production Flow Shop Environments

Harlem M.M. Villadiego, José Elías C. Arroyo, and André Gustavo dos Santos

Department of Computer Science
Universidade Federal de Viçosa, UFV
Viçosa, MG, Brazil
harlem.villadiego@ufv.br, {jarroyo,andre}@dpi.ufv.br

Abstract. In this work, we consider the integrated lot-sizing and sequencing problem in a permutation flow shop with machine sequence-dependent setups. The problem is to determine the lot sizes and the production sequence in each period of a planning horizon such that the customer demands must be met and the capacity of the machines must be respected. The objective is to determine the sum of the setup costs, the production costs and the inventory costs over the planning horizon. Due to the complexity of the problem, we propose a heuristic based on Iterated Greedy metaheuristic which uses sequencing and lot-sizing decisions. The proposed method is compared against the best heuristics available in the literature in a large set of problem instances. Comprehensive computational and statistical analyses are carried out in order to validate the performance of the proposed heuristic.

Keywords: Lot-sizing, flow shop scheduling, metaheuristics.

1 Introduction

Production planning is the most important issues of production industries since it is the activity that allows to coordinate and to conduct all operations in a production process in order to meet the commitments made to customers of the company. According to [1], production planning involves lot-sizing and scheduling decisions. A lot-sizing indicates the quantity of a product (or lot of a product) manufactured on a machine continuously without interruption. The scheduling is responsible for establishing the sequence of products to be produced determining their start and completion time, in a time period.

In the literature several studies address the problem of lot sizing and scheduling jointly. Problems that have features such as multi-product, capacitated and sequence-dependent setup time are more common. Barany et al. [2] proposed strong formulations for multi-product capacitated lot sizing problem (CLSP) obtaining good results in a single machine. Almada-Lobo et al. [3] proposed a heuristic method with exact formulations for the CLSP considering sequence-dependent setup times. Almada-Lobo et al. [4] also investigated the CLPS with

C. Blum and G. Ochoa (Eds.): EvoCOP 2014, LNCS 8600, pp. 61–72, 2014.

sequence-dependent setup times. These authors proposed a Variable Neighbour-hood Search heuristic and the results were better when compared to the heuristic proposed by Almada-Lobo et al. [3]. A literature review for the CLSP can be found in [5]. Toledo et al. [6] investigated the multi-level capacitated lot-sizing problem (MCLSP) in parallel machines and proposed a multi-population ge-netic algorithm with mathematical programming techniques. The MCLSP is considered when is a dependency between the production of various products in different levels of production. The production environment in the MCLSP can be: single machine, parallel machine, flow shop, among others. Models and algorithms for the MCLSP are discussed in Karimi et al. [7]. A discussion of lot-sizing and scheduling with sequence-dependent setup times is found in Clark and Clark [8]. Zhu and Wilhelm [9] presented a review of lot-sizing and scheduling problems and conclude that researches on flow shop environment are scarce.

In this paper, we investigate the simultaneous lot-sizing and scheduling prob-lem in flow shop (LSSPFS) production environment. The first work for the LSSPFS was presented by Sikora et al. [10]. These authors proposed a heuristic method based on two non-integrated approaches: one for lot-sizing and other for the scheduling. For the same problem, Sikora [11] proposed a genetic algo-rithm which performs better than the heuristic method proposed by Sikora et al. [10]. Ponnambalam et al. [12] proposed a hybrid heuristic that combines ge-netic algorithm and simulated annealing. In this heuristic, the genetic algorithm attempts to solve the lot-sizing while the simulated annealing solve the schedul-ing problem. The results obtained by this hybrid heuristic are better than those presented by Sikora [11]. Smith et al. [13] propose a mathematical model for the LSSPFS. Due to the large number of decision variables that presents in the mathematical model, this is only feasible for smaller instances.

Mohammadi et al. [14] presents a mixed-integer programming (MIP) model for the LSSPFS with machines sequence-dependent setups. In this problem, de-mands must be met, the capacity of the machines must be respected, and setups are preserved between periods of the planning horizon. Mohammadi et al. [14] also developed two lower bounds, two heuristic based on the "rolling Horizon" strategy and two heuristic based on "relax and fix" strategy. The authors claim that the heuristics "relax and fix" achieved better results. Mohammadi et al. [15] extend the work done by Mohammadi et al. [14]. They used the same MIP model, the same lower bounds and proposed some improvements in the heuris-tics. Mohammadi et al. [16] considered the same MIP model and lower bounds of Mohammadi et al. [14], however they introduced two new heuristics based on "rolling horizon" strategy and on the classical NEH heuristic [17], respectively. Belo Filho et al. [18] proposed an Asynchronous Teams (A_Times) heuristic which is compared with the methods proposed by Mohammadi et al. [15]. These authors showed that the A_Times heuristic provides the best results. Rameza-nian et al. [19] propose a new and more efficient MIP model for LSSPFS. This model has a smaller number of decision variables and constraints. Besides the new model, these authors proposed also two heuristics based on the "rolling

horizon" strategy. The heuristics outperformed the methods proposed by Mohammadi et al. [16].

In this work, we propose a Iterated Greedy (IG) heuristic to solve the same
LSSPFS problem which is also addressed in [16], [18] and [19]. IG is a simple and
efficient metaheuristic proposed by Ruiz et al. [20] for the flow shop scheduling
problem. Our IG heuristic tries to find the best production sequence (processing
order of products) in each period by minimizing the setup costs. For an obtained
production sequence, the lot sizes are determined by a backward and forward
improvement method, similar to that proposed by Shim et al. [21]. The proposed
IG heuristic is compared against the two best heuristics available in the literature: the A_times heuristic proposed by Belo Filho et al. [18] and the Rolling
Horizon (RH) based heuristic proposed by Ramezanian et al. [19].

2   Problem Description

This paper is considered the integrated problem of lot sizing and scheduling.
Each lot is a specific quantity of a product, which must be processed in a production environment of type flow shop, where machines are limited in capacity
and arranged in series in a finite planning horizon and divided into periods.
There a certain demand for each product in each period, which must always
be attended without delay. Therefore, must determine the amount to produce
(lot sizing) of each product, as well the order in which they will be processed
on the machines in each period. The objective function (f) of the problem is
to minimize the sum of the production cost, sequence-dependent setup and inventory costs. This problem is well stated by Mohammadi et al [14]. The main
assumptions of the problem are summarized as fallows:

- Lot-sizing and scheduling are made simultaneously.
- The finite planning horizon is divided into T periods.
- At the beginning of the planning horizon machines are set for a specific
  product.
- Several products can be manufactured in each period.
- Each machine has a processing capacity.
- At any time, each machine can process at most one product.
- At any time, each product can be processed on at most one machine.
- The setup of a machine must be completed in a period.
- Sequence-dependent setup costs and times occur for product changes on
  machinery.
- The triangle inequality with respect to the setup cost and time holds holds,
  i.e., it is never faster to change over from one product to another by means
  of a third product.
- External demands for the finished products are known.
- Shortage is not permitted.
- A component cannot be produced in a period until the production of its
  required components is finished. In other words, production at a level can
  only be started if a sufficient amount of the required products from the
  previous level are available; this is called vertical interaction.

3   Iterated Greedy Heuristic for the LSSPFS

In this section, we develop an Iterated Greedy (IG) heuristic to tackle the LSSPFS problem. The proposed IG tries to improve iteratively a solution through four different stages: destruction-construction, local search, lot-sizing improvement and the acceptance criterion. IG starts by generating an initial solution by means of a constructive procedure. Then, it applies iteratively the above mentioned four steps process until a predefined stopping condition is satisfied. The stages destruction-construction and local search are used to generate a good production sequence. The lot-sizing improvement stage is applied to the best production sequence obtained by the local search in order to improve the lot-sizing quality. The new current solution is chosen between the current solution and the solution obtained from the lot-sizing improvement stage is chosen at each iteration (acceptance criterion).

The general scheme of the proposed IG heuristic is presented in Algorithm 1. It starts by generating an initial solution (line 1). If a unfeasible solution is built, we use the MIP model proposed by [19] to obtain a valid solution (lines 2 to 4). The iterations of the IG are computed in lines 6 to 16 until the stopping criterion is satisfied (maximum number of iterations). In lines 7 and 8 are executed the destruction-construction and local search procedures, respectively. The lot-sizing improvement is run in line 9. Lines 10 through 14 is tested the acceptance criterion. The best obtained solution is returned by the algorithm (line 17).

Algorithm 1. IG(d, It_ls, It_bf, It_IG)

---

1. $SOL :=$ Construction_Initial_Solution();
2. **if** Infeasible_Solution(SOL) **then**
3. $SOL :=$ MIP(SOL);
4. **end if**
5. $SOL^* := SOL$;
6. **while** $iterations \leq It\_IG$ **do**
7. $SOL :=$ Destruction_Construction(SOL, d);
8. $SOL :=$ Local_Search(SOL, It_ls);
9. $SOL :=$ Lot-Sizing_Improvement(SOL, It_bf);
10.   **if** $f(SOL) < f(SOL^*)$ **then**
11. $SOL^* := SOL$;
12.   **else**
13. $SOL := SOL^*$;
14.   **end if**
15. $iterations := iterations + 1$;
16. **end while**
17. **return** SOL^*;

In next subsections we detail the representation of a solution, the construction of the initial solution and the four stages of the proposed IG heuristic.

3.1   Solution Representation

In the IG heuristic, a solution is represented by T (number of periods) matrices of order $3 \times N$ (where, N is number of products). The first row of each matrix represents the production sequence, the second row stores the quantity produced of each product and the next row stores the inventories of the products. Figure 1 shows an example of a solution for two periods (T_1 and T_2) and three products (J_1, J_2 and J_3). The production sequence in the first period is (J_3, J_2, J_1), and in the second period is (J_1, J_2, J_3). LS_{jt} and I_{jt} represent, respectively, the lot sizing and inventory for each product j in period t.

	T_1			T_2		
production sequence	J_3	J_2	J_1	J_1	J_2	J_3
Lot Sizing	LS_{31}	LS_{21}	LS_{11}	LS_{12}	LS_{22}	LS_{32}
Inventory	I_{31}	I_{21}	I_{11}	I_{12}	I_{22}	I_{32}

Fig. 1. Solution Representation

3.2   Construction of an Initial Solution

The construction of an initial solution is made in two phases. In the first phase, the production sequence for each period is constructed. In the second phase, from the obtained production sequence, the lot sizes are determined.

The production sequences are generated applying the Total Heuristic (TH) proposed by Simons et al. [22] for the flow shop scheduling problem with sequence dependent setup times. In this work, we minimize the setup costs over all the machines. In TH heuristics, from period 2 is taken into account the last product processed in the prior period to maintain preservation of the machines between periods (setup carryover).

Considering the production sequences previously obtained in the first phase, the lot sizes are determined in backward direction, that is, from the last period to the first period [21]. If the total cumulative remaining demand is less than or equal to the available capacity, the lost sizes are set to the corresponding demands. Otherwise, the lot sizes are set to the corresponding demands from the last product in the sequence considering the available capacity. When the demand of a determined period violates the available capacity, this demand is moved to preceding period. In some cases, we can to obtain an unfeasible solution. When this occurs, we execute the MIP model proposed by Ramezanian et al. [19] to determine the lot sizes. This model considers the production sequence obtained in the first phase of the construction procedure.

3.3   Destruction and Construction Procedures

The destruction procedure is applied to each production sequence π of N product. This procedure removes randomly d different product of π. The results of this procedure are two sub-sequences, the first being the partial sequence π_D with $N - d$ product and the second being a sequence of removed product, which we denote as π_R. π_R contains the products that have to be reinserted into π_D to yield a complete sequence. The Construction procedure starts with the sub-sequence π_D and performs d steps in which the products in π_R are reinserted into π_D. The process start inserting the first product of π_R, into all possible $N - d + 1$ positions of π_D. The best position for this product in π_D sequence is the one that yields the smallest setup cost. This process is iterated until π_R is empty. The destruction and construction procedures are applied to the production sequence of each period.

3.4   Local Search

The sequences returned by Destruction and Construction procedures are im-proved by a Local Search procedure. This procedure is based on the interchange and insertion movements. This Local Search is a variant of the Variable Neigh-borhood Search method to solve the permutation flow shop proposed by Tasge-tiren et al. [23]. The interchange movement randomly selects two products in the sequence and exchanges their positions. The insertion movement removes a ran-dom product from its original position and inserts in all possible position. The objective function to be minimized in the Local Search is also the setup costs (f_{sc}). The Local Search procedure is detailed in Algorithm 2. This algorithm receives as parameters a production sequence (π) and the maximum number of iterations (It_ls) which is used as stop condition.

Algorithm 2. $Local\_Search(It\_ls, \pi)$

---

1. **for** $l := 1$ **to** It_ls **do**
2.   improved := true
3.   **while** improved = true **do**
4. $\phi := \pi$
5.   **if** improved = true **then**
6. $\phi :=$ sequence obtained by swapping two adjacent products randomly;
7.   **else**
8. $\phi :=$ the best sequence obtained by inserting an product randomly in all possible positions;
9.   **end if**
10.   **if** $f_{SC}(\phi) < f_{SC}(\pi)$ **then**
11. $\pi := \phi$;
12.   improved := true
13.   **else**
14.   improved := false
15.   **end if**
16.   **end while**
17. **end for**
18. **return** π;

3.5   Lot-sizing Improvement

The Lot-sizing improvement consists of two methods: backward and forward improvement methods. These methods are similar to those proposed by Shim et al. [21]. The backward method, shifts a certain amount of a product (say, J) from period T_i to an earlier period T_{i-1}. Depending on the available capacity in period T_{i-1}, we can only move the entire lot of product J. Before a lot is moved, we check the period T_{i-1} for the available capacity. If product J is already produced in period T_{i-1}, no new set-up is needed and all the slack capacity available in period T_{i-1} can be used to produce more of product J. Due that the lot-splitting is not considered, the idea is to reduce the production and setup cost. The forward method, is the inverse of backward method. Here, a certain amount of a product J in period T_i is shifted to a later period T_{i+1}. The amount of production shifted (whole lot or a fraction) depends on available capacity in period T_{i+1}. Note that the main purpose of the forward improvement is to reduce the inventory holding cost. In both methods, the product J and the period T_i are randomly selected. The backward and forward methods are repeated It_bf times and the best solution found is returned. Figure 2. (a) illustrates an example of backward method for two periods. In this example the quantities produced of product J_2 in period T_2 (LS_{22}) is moved to the period T_1. This reduce production and setup costs in period T_2. The Figure 2. (b) illustrates an example of forward method, where part of the production of product J_2 in the period T_1 (I_{21}) is moved to the next period (T_2). This movement reduces inventory costs in period T_1.

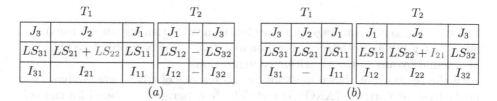

Fig. 2. Backward and Forward methods

4   Computational Experiments

In this section we present the computational test made to analyze the performance of the proposed IG heuristic. All tests were run on an $Intel(R)$ $Xeon(R)$ CPU $X5650$ $2.67GHz$ with $48GB$ of RAM. The results obtained by IG heuristic are compared with the results of A_Times and RH heuristics proposed by Belo Filho et al. [18] and Ramezanian et al. [19], respectively. These heuristics were re-implemented according to the original papers. In the A_times implementation, we use the MIP model proposed by Ramezanian et al. [24]. All algorithms were coded in $C++$, using classes and libraries of IBM ILOG CPLEX 12.4 and CPLEX Concert Technology.

In order to evaluate the performance of the proposed heuristic, we generated 300 instances of the problem, by combining different number of products, machines and periods: (N, M, T). The sizes of the instances vary in the range of $(3, 3, 3)$ to $(65, 65, 65)$. These instances were generated according to the test data proposed by Mohammadi et al. [14]. For each size (N, M, T), five instances were generated. Table 1 shows the sizes of the generated instances.

4.1 Calibration of the IG Heuristic

For the IG heuristic, four parameters were adjusted: destruction parameter (d), number of iterations of local search (It_ls), number of iterations of the lot-sizing improvement (It_bf) and the total iterations of IG heuristic (It_IG). For each parameter, three levels of values were tested: $d = 0.2N, 0.4N, 0.6N$; $It\_ls = 3, 5, 7$; $It\_bf = 4, 6, 8$ and $It\_IG = 100T, 150T, 200T$. Note that, by combining the values of the four parameters we have 81 configurations of the IG heuristic.

To determine the best combination of the parameters, a computational experiment is carried out. We employ a Design of Experiments approach where each parameter is a controlled factor. All the 81 configurations were tested in a full factorial experimental design [25]. For each configuration of the values, the algorithm was run ten times to solve each test instance. As response variable for the experiments, we use the Relative Percentage Deviation (RPD) which is computed in the following way:

$$RPD\% = 100 \times \frac{f_{method} - f_{best}}{f_{best}} \tag{1}$$

where f_{method} is the average objective function value (among the ten runs) obtained by a given algorithm configuration and f_{best} is the best objective function value found by any of the algorithm configurations.

All the results (not shown in detail due to reasons of space) are analyzed by the Analysis of Variance (ANOVA) test. The best parameters found for the IG heuristic were: $d = 0.6N$, $It\_ls = 5$, $It\_bf = 8$ and $It\_IG = 150T$.

4.2 Results and Comparisons

The obtained results by the three heuristics, IG, RH and A_Times, are compared by using the RPD (equation (1)). To solve each test problem, each heuristic was run 10 times. Table 2 shows the average RPD for the three heuristics for the last twenty large instances. The average best results are in boldface. We can see that the IG heuristic shows a very good performance and provides the best RPD values in thirteen of twenty types of large instances.

In order to validate the results obtained by the three heuristic and check if the observed differences are statistically significant between them, an ANOVA test was performed, where the RPD is considered as non-controllable factor. In this analysis, we use all the 300 instances, from smaller instance $(3, 3, 3)$ to

Table 1. Size of the used instances

Small	Medium	Large
(3, 3, 3)	(20, 15, 15)	(45, 40, 40)
(5, 3, 3)	(15, 20, 15)	(40, 45, 40)
(3, 5, 3)	(15, 15, 20)	(40, 40, 45)
(3, 3, 5)	(20, 20, 20)	(45, 45, 45)
(5, 5, 5)	(25, 20, 20)	(50, 45, 45)
(7, 5, 5)	(20, 25, 20)	(45, 50, 45)
(5, 7, 5)	(20, 20, 25)	(45, 45, 50)
(5, 5, 7)	(25, 25, 25)	(50, 50, 50)
(7, 7, 7)	(30, 25, 25)	(55, 50, 50)
(10, 5, 5)	(25, 30, 25)	(50, 55, 50)
(5, 10, 5)	(25, 25, 30)	(50, 50, 55)
(5, 5, 10)	(30, 30, 30)	(55, 55, 55)
(10, 7, 7)	(35, 30, 30)	(60, 55, 55)
(7, 10, 7)	(30, 35, 30)	(55, 60, 55)
(7, 7, 10)	(30, 30, 35)	(55, 55, 55)
(10, 10, 10)	(35, 35, 35)	(60, 60, 60)
(15, 10, 10)	(40, 35, 35)	(65, 60, 60)
(10, 15, 10)	(35, 40, 35)	(60, 65, 60)
(10, 10, 15)	(35, 35, 40)	(60, 60, 65)
(15, 15, 15)	(40, 40, 40)	(65, 65, 65)

Table 2. Average RPD for the heuristics

Instance	IG	A_Times	RH
(45, 40, 40)	**0.2380**	1.5631	2.5274
(40, 45, 40)	**0.1332**	1.1488	1.3230
(40, 40, 45)	0.3519	1.7013	**0.2512**
(45, 45, 45)	0.3600	0.0896	**0.0861**
(50, 45, 45)	**0.2134**	2.8277	3.1790
(45, 50, 45)	**0.1926**	0.8997	2.9597
(45, 45, 50)	0.8802	**0.2253**	3.2108
(50, 50, 50)	1.6614	**0.1936**	2.0947
(55, 50, 50)	0.3182	0.3391	**0.1783**
(50, 55, 50)	**0.0571**	0.1800	0.1536
(50, 50, 55)	**0.1793**	0.2617	2.4745
(55, 55, 55)	**0.1980**	1.6696	1.4926
(60, 55, 55)	**0.0901**	0.1642	0.1108
(55, 60, 55)	**0.1235**	0.5582	0.7605
(55, 55, 55)	0.2219	**0.0854**	1.2726
(60, 60, 60)	**0.1936**	1.5123	3.4994
(65, 60, 60)	**0.1489**	1.5879	2.2476
(60, 65, 60)	0.4144	**0.0725**	0.5391
(60, 60, 65)	**0.0941**	1.7381	0.1630
(65, 65, 65)	**0.1078**	0.5641	0.7282

the largest $(65, 65, 65)$. Since the p-value found through the ANOVA is 0.0005, which is smaller than 0.05, there is a statistical significant difference between them heuristic with a confidence level of 95%.

The ANOVA does not specify which heuristics are different. So, we use the Multiple Comparisons test for comparing each pair of means with a 95% confidence level. Table 3 shows the result of this test. In this Table, the "Difference" column displays the sample mean of the first algorithm minus that of the second.

The "+/- Limits" column shows an uncertainty interval for the difference. The heuristics pairs, for which the absolute value of the difference exceeds the limit, are statistically different at the selected confidence level (95%). This difference is indicated by an (*) in the "Significant" column. In this Table, we can see that two pairs of heuristics have significant statistical difference. That is, a statistical significant difference between the pairs of heuristic $IG - RH$ and $IG - A\_Times$. There is not statistical significant difference between the pair of heuristic $RH - A\_Times$. The same analysis can be displayed in the Figure 3a. This Figure shows the means plot and Tukey's Honestly Significant Difference (HSD) intervals at 95% confidence level from the Multiple Comparisons test. Since the interval for the IG heuristics does not overlap any of the other intervals, the mean of IG is significantly different of the other two heuristic.

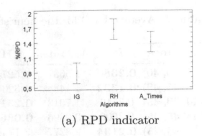

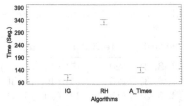

(a) RPD indicator   (b) Average computational times

Fig. 3. Means plot and Tukey's HSD 95% confidence intervals for the evaluated algorithms

Table 3. Multiple Range Tests for RPD

Contrast	Significant	Difference	+/- Limits
IG - RH	*	-0.8744	0.411
IG - A_Times	*	-0,6257	0.411
RH - A_Times		0.2487	0.411

4.3 Analysis of the Computational Times

The experiments described in 4.2 were made considering the quality of the solutions with respect to the objective function value. However, it is also important to consider the performance of the heuristics on the CPU time. We execute an ANOVA test in which the response variable under study is the average computational time presented by each heuristic. Since the p-value found through the ANOVA is 0.0000, which is smaller than 0.05, there is a statistically significant difference between all the average computational time of the heuristic with a confidence level of 95%. Table 4 shows the result of the Multiple Comparisons test for comparing each pair of average computational time. In this Table, we can see that there are significant differences between all pairs of heuristics. These results indicate that, with a confidence level of 95%, the IG heuristic spends less computational time.

Table 4. Multiple Range Tests for average time

Contrast	Significant	Difference	+/- Limits
IG - RH	*	-222.2	21.59
IG - A_Times	*	-29.0	21.59
RH - A_Times	*	193.2	21.59

In Figure 3b we can see that all average computational times of the heuristics are significantly different with a confidence level of 95 %. Note that the IG heuristic has the lowest average computational time. The Figure shows that the average computational time of *IG*, *RH* and *A_Time* are 107.2, 329.3 and 136.2 seconds, respectively.

5   Conclusion

The integrated lot-sizing and scheduling problem in flow shop environment has been discussed in this paper. To determine the lot sizes and the production sequence in each period of a planning horizon, we suggested a IG heuristic. In order to evaluate the performance of the proposed heuristic, new large instances were generated according to Mohammadi et al. [14]. We have obtained the best combination of parameters for the IG heuristic by performing a design of experiments approach. After an exhaustive computational and statistical analysis we can conclude that the proposed IG heuristic shows an excellent performance overcoming the two heuristics, A_Times and RH, in a comprehensive benchmark set of instances. Future research directions involve the consideration of new approaches for lot sizing decision, and apply the IG heuristic to simultaneous lot-sizing and scheduling problem in another machine environment.

References

1. Pochet, Y., Wolsey, L.A.: Production planning by mixed integer programming. Springer (2006)
2. Barany, I., Van Roy, T.J., Wolsey, L.A.: Strong formulations for multi-item capacitated lot sizing. Management Science 30(10), 1255–1261 (1984)
3. Almada-Lobo, B., Klabjan, D., Antónia Carravilla, M., Oliveira, J.F.: Single machine multi-product capacitated lot sizing with sequence-dependent setups. International Journal of Production Research 45(20), 4873–4894 (2007)
4. Almada-Lobo, B., James, R.J.: Neighbourhood search meta-heuristics for capacitated lot-sizing with sequence-dependent setups. International Journal of Production Research 48(3), 861–878 (2010)
5. Drexl, A., Kimms, A.: Lot sizing and scheduling survey and extensions. European Journal of Operational Research 99(2), 221–235 (1997)
6. Toledo, C., Frana, P., Morabito, R., Kimms, A.: Multi-population genetic algorithm to solve the synchronized and integrated two-level lot sizing and scheduling problem. International Journal of Production Research 47(11), 3097–3119 (2009)
7. Karimi, B., Ghomi, S.F., Wilson, J.: The capacitated lot sizing problem: A review of models and algorithms. Omega 31(5), 365–378 (2003)
8. Clark, A.R., Clark, S.J.: Rolling-horizon lot-sizing when set-up times are sequence-dependent. International Journal of Production Research 38(10), 2287–2307 (2000)
9. Zhu, X., Wilhelm, W.E.: Scheduling and lot sizing with sequence-dependent setup: A literature review. IIE Transactions 38(11), 987–1007 (2006)
10. Sikora, R., Chhajed, D., Shaw, M.J.: Integrating the lot-sizing and sequencing decisions for scheduling a capacitated flow line. Computers & Industrial Engineering 30(4), 659–679 (1996)
11. Sikora, R.: A genetic algorithm for integrating lot-sizing and sequencing in scheduling a capacitated flow line. Computers & Industrial Engineering 30(4), 969–981 (1996)
12. Ponnambalam, S., Reddy, M.: A ga-sa multiobjective hybrid search algorithm for integrating lot sizing and sequencing in flow-line scheduling. The International Journal of Advanced Manufacturing Technology 21(2), 126–137 (2003)

13. Smith-Daniels, V., Ritzman, L.P.: A model for lot sizing and sequencing in process industries. The International Journal of Production Research 26(4), 647–674 (1988)
14. Mohammadi, M., Ghomi, S.F., Karimi, B., Torabi, S.A.: Rolling-horizon and fix-and-relax heuristics for the multi-product multi-level capacitated lotsizing problem with sequence-dependent setups. Journal of Intelligent Manufacturing 21(4), 501–510 (2010)
15. Mohammadi, M., Ghomi, S.F., Karimi, B., Torabi, S.: Mip-based heuristics for lot-sizing in capacitated pure flow shop with sequence-dependent setups. International Journal of Production Research 48(10), 2957–2973 (2010)
16. Mohammadi, M., Torabi, S.A., Ghomi, S.F., Karimi, B.: A new algorithmic approach for capacitated lot-sizing problem in flow shops with sequence-dependent setups. The International Journal of Advanced Manufacturing Technology 49(1-4), 201–211 (2010)
17. Nawaz, M., Enscore, E.E., Ham, I.: A heuristic algorithm for the m-machine, n-job flow-shop sequencing problem. Omega 11(1), 91–95 (1983)
18. Belo Filho, M.A.F., dos Santos, M.O., de Meneses, C.N.: Dimensionamento e sequenciamento de lotes para uma linha de producao flowshop: Métodos de solução (2012)
19. Ramezanian, R., Saidi-Mehrabad, M., Teimoury, E.: A mathematical model for integrating lot-sizing and scheduling problem in capacitated flow shop environments. The International Journal of Advanced Manufacturing Technology, 1–15 (2012)
20. Ruiz, R., Stützle, T.: An iterated greedy heuristic for the sequence dependent setup times flowshop problem with makespan and weighted tardiness objectives. European Journal of Operational Research 187(3), 1143–1159 (2008)
21. Shim, I.S., Kim, H.C., Doh, H.H., Lee, D.H.: A two-stage heuristic for single machine capacitated lot-sizing and scheduling with sequence-dependent setup costs. Computers & Industrial Engineering 61(4), 920–929 (2011)
22. Simons Jr., J.: Heuristics in flow shop scheduling with sequence dependent setup times. Omega 20(2), 215–225 (1992)
23. Tasgetiren, M.F., Liang, Y.-C., Sevkli, M., Gencyilmaz, G.: A particle swarm optimization algorithm for makespan and total flowtime minimization in the permutation flowshop sequencing problem. European Journal of Operational Research 177(3), 1930–1947 (2007)
24. Ramezanian, R., Saidi-Mehrabad, M.: Hybrid simulated annealing and mip-based heuristics for stochastic lot-sizing and scheduling problem in capacitated multi-stage production system. Applied Mathematical Modelling 37(7), 5134–5147 (2013)
25. Montgomery, D.G.: Design and Analysis of Experiments. Wiley, New York (2001)

Balancing Bicycle Sharing Systems: An Approach for the Dynamic Case*

Christian Kloimüllner, Petrina Papazek, Bin Hu, and Günther R. Raidl**

Institute of Computer Graphics and Algorithms
Vienna University of Technology
Favoritenstraße 9–11/1861, 1040 Vienna, Austria
{kloimuellner,papazek,hu,raidl}@ads.tuwien.ac.at

Abstract. Operators of public bicycle sharing systems (BSSs) have to regularly redistribute bikes across their stations in order to avoid them getting overly full or empty. We consider the dynamic case where this is done while the system is in use. There are two main objectives: On the one hand it is desirable to reach particular target fill levels at the end of the process so that the stations are likely to meet user demands for the upcoming day(s). On the other hand operators also want to prevent stations from running empty or full during the rebalancing process which would lead to unsatisfied customers. We extend our previous work on the static variant of the problem by introducing an efficient way to model the dynamic case as well as adapting our previous greedy and PILOT construction heuristic, variable neighborhood search and GRASP. Computational experiments are performed on instances based on real-world data from Citybike Wien, a BSS operator in Vienna, where the model for user demands is derived from historical data.

1 Introduction

Bicycle Sharing Systems (BSSs) are evolving in large cities all over the world. They offer various advantages regarding urban development, attractiveness for citizens, reduce individual motorized traffic and complement public transport. Furthermore, BSSs also contribute to public health by encouraging people to do sports [1]. A BSS consists of multiple bike stations distributed over various strategically favorable positions in the city. A registered user is allowed to rent a bike at a station and return it later at another station. Due to miscellaneous factors such as altitude of stations, demographic characteristics, or nearby public transport stops, some stations tend to run empty whereas others tend to get full. In case of an empty station, customers are not able to rent bikes while in case of a full station, customers cannot return their bikes. Therefore, BSS operators need to redistribute bikes among stations on a regular basis to avoid or at least

---

* This work is supported by the Austrian Research Promotion Agency (FFG), contract 831740.
** The authors thank Matthias Prandtstetter, Andrea Rendl and Markus Straub from the Austrian Institute of Technology (AIT) for the collaboration in this project.

C. Blum and G. Ochoa (Eds.): EvoCOP 2014, LNCS 8600, pp. 73–84, 2014.

minimize customer dissatisfaction. Usually this task is done by a vehicle fleet that picks up bikes from stations with excesses of bikes and delivers them to stations with deficits.

When trying to approach our definition of Balancing Bicycle Sharing System (BBSS) problem, the goal is to find a route for every vehicle with corresponding loading instructions, respectively, so that the system is brought to a balanced state and is able to fulfill user demands as much as possible. So far, almost all of our recent work considered only the static variant where it is assumed that the rebalancing process is done while the system is not in use [2–5]. This is a useful approach for BSSs which, e.g., do not operate overnight, and is also practical for strategic planning to reach desired fill levels in the long term as it depicts a simplification to the problem. In this work we extend our previous algorithms for the static case towards the dynamic scenario where we take user demands over time into account, and try to reduce unfulfilled demands during the rebalancing process as well as reaching target fill levels for stations at the end. We propose an efficient way to model and simulate these dynamics as well as adapt a greedy and PILOT construction heuristic, Variable Neighborhood Search (VNS) and GRASP accordingly.

2   Related Work

BBSS can be regarded as a special variant of the capacitated single commodity split pickup and delivery vehicle routing problem. Particular features are that we allow multiple visits of stations, consider heterogeneous vehicles, and the possibility of loading or unloading an arbitrary number of bikes.

Most related approaches address the static variant of BBSS and apply Mixed Integer Programming (MIP) techniques. A direct comparison of existing works is difficult as most of them consider different problem characteristics. Chemla et al. [6] propose a branch-and-cut algorithm on a relaxed MIP model in conjunction with a tabu search for obtaining upper bounds. However, they assume only a single vehicle and reaching the target fill levels as a hard constraint. Benchimol et al. [7] also consider a single vehicle and balance as a hard constraint and propose approximation algorithms. Raviv et al. [8] use MIP approaches with a convex penalty objective function to minimize user dissatisfaction and tour lengths for multiple vehicles. However, they ignore the number of loading operations. In our recent works we developed several metaheuristic approaches which scale well for large instances [3–5]. In particular, we introduced a Greedy Construction Heuristic (GCH) and a VNS approach in [3], a PILOT construction heuristic and GRASP in [5]. Different strategies for finding meaningful loading instructions for candidate routes including optimal ones were studied in [2]. In [4] we refined our concepts and provided more extensive computational analysis. Di Gaspero et al. [9, 10] investigate Constraint Programming (CP) approaches in conjunction with ant colony optimization, a smart branching strategy, and large neighborhood search. They test on the same BBSS variant as we do, but could not outperform our VNS approach from [3].

Concerning the dynamic BBSS scenario, there exist only few MIP approaches so far. Contardo et al. [11] present an approach utilizing Dantzig-Wolfe and Benders decomposition. They obtain upper and lower bounds for instances with up to 100 stations, but face significant gaps. Unlike our problem definition, they focus exclusively on fulfilling user demands but do not consider target fill levels. Schuijbroek et al. [12] apply a clustering-first route-second strategy. A clustered MIP heuristic, or alternatively, a CP approach handle the routing problems. In contrast to our work they define intervals for fulfilling user demands. These are considered as hard constraints whereas we try to minimize as many unfulfilled demands as possible. Additionally, they do not consider target fill levels. Chemla et al. [13] present a theoretical framework to estimate the vehicles' impacts on the system and propose heuristic approaches for a single vehicle. Besides, they suggest a pricing strategy to encourage users to return bikes at stations which tend to run empty soon. Pfrommer et al. [14] investigate a heuristic for planning tours with multiple vehicles and also suggest a dynamic pricing strategy. They periodically recompute truck tours and dynamic prices while the system is active and test with a simulation based on historic data.

Other related works examine strategic planning aspects of BSSs such as location and network design [15, 16] or system characteristics and usage patterns [17]. However, these aspects are not within the scope of this work.

3   Problem Definition

We consider the dynamic scenario of BBSS, referred to as *DBBSS*, where rebalancing is done while we simulate system usage by considering expected cumulated user demands. In addition to the input data for the static problem variant we particularly consider expected user demands from a prediction model.

For the BSS infrastructure we are given a complete directed graph $G = (V \cup \{0\}, A)$ where node set V represents rental stations, node 0 the vehicles' depot, and arc set A the fastest connection between all nodes. Each arc $(u, v) \in A$ is assigned a weight corresponding to the travel time $t_{u,v} > 0$ (including average times for loading and unloading actions). Each station $v \in V$ has a capacity $C_v \geq 0$ denoting the total number of bike slots. The initial fill level p_v is the number of available bikes at the beginning of the rebalancing while the target fill level q_v states the desired number of bikes at the end of the rebalancing. For the rebalancing procedure we are given a fleet of vehicles $L = \{1, \ldots, |L|\}$ where each vehicle $l \in L$ has a capacity $Z_l > 0$. Finally, let $\hat{t}_{\max}$ be the time budget within a vehicle has to finish its route which starts and ends at the depot 0.

Regarding user demands over time we assume the *expected cumulated demand* $\mu_v(t) \in \mathbb{R}$ occurring at each station $v \in V$ from the beginning of the rebalancing process until time t, $0 \leq t \leq \hat{t}_{\max}$ to be given as an essentially arbitrary function. The cumulated demand is calculated by subtracting the expected number of bikes to be returned from the expected number of bikes to be rent over the respective time period. An example of a demand function is shown in Figure 1. Note that we display $p_v - \mu_v(t)$ as the dash-dotted line in order to highlight the

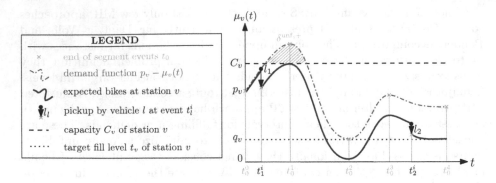

Fig. 1. Example of a demand function and two pickup events

area where unfulfilled demands occur. Thus, a positive slope of $\mu_v(t)$ indicates that more users are expected to rent bikes than to return them in the time period t, and vice versa. Demands are always fulfilled immediately as far as possible, i.e., bikes or parking slots are available. Unfulfilled demands cannot be fulfilled later and are penalized in the objective function. Let $\hat{\delta}_v^{\mathrm{unf},-}$ denote the total amount of *unfulfilled bike demands* for station $v \in V$, i.e., the number of users who want to rent a bike but are not able to at the desired station. Analogously, let $\hat{\delta}_v^{\mathrm{unf},+}$ refer to the total number of *unfulfilled slot demands*, i.e., the amount of users who cannot return bikes as the station is already full.

A solution to DBBSS consists of a route for every vehicle and corresponding loading instructions for every stop at a station. A route of length ρ_l is defined as an ordered, arbitrarily long sequence of stations $r_l = (r_l^1, \ldots, r_l^{\rho_l})$, $r_l^i \in V$ where the depot is assumed to be implicitly added as start and end node. The loading instruction for vehicle $l \in L$ during the i-th stop at station $v \in V$ is denoted as $y_{l,v}^i$. Positive values for $y_{l,v}^i$ denote the corresponding number of bikes to be picked up, negative values denote deliveries. Feasible solutions must fulfill the following conditions. For any station, its fill level (i.e., the number of currently available bikes) must always lie between 0 and its capacity C_v. For any vehicle $l \in L$ the load may never exceed its capacity, i.e., $b_l \leq Z_l$. Moreover, a solution is only feasible, if and only if no route's total travel time, denoted by t_l, exceeds the time budget, and additionally, every vehicle must return empty to the depot 0.

The goal is to find a route for each vehicle with corresponding loading instructions such that the following objective function is minimized:

$$f(r,y) = \omega^{\mathrm{unf}} \sum_{v \in V} (\hat{\delta}_v^{\mathrm{unf},-} + \hat{\delta}_v^{\mathrm{unf},+}) + \omega^{\mathrm{bal}} \sum_{v \in V} |q_v - p_v|$$

$$+ \omega^{\mathrm{load}} \sum_{l \in L} \sum_{i=1}^{\rho_l} |y_{l,r_l^i}^i| + \omega^{\mathrm{time}} \sum_{l \in L} t_l \qquad (1)$$

Parameters ω^{unf}, ω^{bal}, ω^{load}, and $\omega^{\mathrm{time}} \geq 0$ are used for controlling the relative importance of the corresponding term in the objective function. The most important goal is to minimize unfulfilled demands as well as to minimize the deviation from the target fill levels. Secondarily, we also want to keep the total number of loading instructions and the total driving time as small as possible, however, those aspects are considered to be clearly less important.

4   Modeling the Dynamic Scenario

In this section we show how DBBSS can be modeled by calculating dynamic behavior of the system, i.e., considering the user demands, so that it can be approached by metaheuristics.

4.1   Segments and Events

One of our major aims is to avoid a time discretization of the demand functions and corresponding fill level calculations as this would introduce errors and is also time consuming if done in an appropriate resolution. Alternatively, we follow the approach of splitting each cumulated demand function into weakly monotonic segments instead of iterating through all discrete time points. Along with the practically reasonable assumption that the number of segments per station is relatively small, such an approach is much more efficient.

For this purpose, we split function $\mu_v(t)$ into monotonically weakly increasing or decreasing segments. Let $t_0 = (t_0^0, \ldots, t_0^{\rho_0})$ with $t_0^0 = 0$ be an ordered sequence of ρ_0 extreme values of $\mu_v(t)$ so that $\mu_v(t)$ is weakly monotonic for $t \in [t_0^{i-1}, t_0^i]$, $\forall i = 1, \ldots, \rho_0$, see Figure 1. Time t_0^i, $i = 1, \ldots, \rho_0$, refers to the end of the i-th weakly monotonic segment. In general, let t_l^i, $\forall l \in L$, $i = 0, \ldots, \rho_l$, be the time when vehicle l performs its i-th stop, i.e.,

$$t_l^i = \begin{cases} 0 & \text{for } i = 0 \\ t_{s_l, r_l^1} & \text{for } i = 1 \text{ if } \rho_l \geq 1 \\ t_l^{i-1} + t_{r_l^{i-1}, r_l^i} & \text{for } i = 2, \ldots, \rho_l \text{ if } \rho_l \geq 2. \end{cases} \tag{2}$$

For each station $v \in V$ we define a data structure which denotes the *series of events* $W_v = \langle (l_1, i_1), \ldots, (l_{|W_v|}, i_{|W_v|}) \rangle$. Each event (l_j, i_j), $j = 1, \ldots, |W_v|$ with $l_j \in \{0\} \cup L$ and $i_j \in \{1, \ldots, \rho_{l_j}\}$ either refers to a *station-visit event*, in which case $l_j \in L$ indicates the corresponding vehicle and i_j the number of its stop, or an *end-of-segment event*, in which case $l_j = 0$ and i_j denotes the respective segment of $\mu_v(t)$. Following the above definitions, the time of event (l_j, i_j) is $t_{l_j}^{i_j}$, and all events in W_v are ordered according to increasing times. Multiple events occurring at the same time are ordered arbitrarily, except that an end-of-segment event always appears last.

4.2   Expected Number of Bikes at Stations

For each station and event we need to derive a fill level considering the cumulated user demand as well as all performed loading or unloading instructions occurred up to this event.

Let $a_{v,j} \in [0, C_v]$ denote the expected number of bikes at station $v \in V$ and event $j = 1, \ldots, |W_v|$ by considering all expected demands fulfilled as far as possible and all pickups and deliveries performed up to and including event j. Note that, as the cumulated user demand is only a forecast model based on historical data, the fill level of every event may also be fractional. Formally, $a_{v,j}$ is calculated as follows:

$$a_{v,j} = \begin{cases} p_v & \text{for } j = 0 \\ \max(\min(a_{v,j-1} - (\mu_v(t_{l_j}^{i_j}) - \mu_v(t_{l_{j-1}}^{i_j-1})), C_v), 0) - y_{l_j}^{i_j} & \text{for } j = 1, \ldots, |W_v|. \end{cases} \quad (3)$$

End-of-segment events are considered for the correct computation of unfulfillable demands. For the ease of notation, the above formula considers them in the same way as *vehicle-visit events*. Since no bikes are delivered or picked up by these events, we define the loading instructions to be $y_0^i = 0$, for $i = 1, \ldots, \rho_0$.

With respect to unfulfilled demands, we distinguish between *unfulfilled bike demands* $\hat{\delta}_v^{\text{unf},-}$ and *unfulfilled slot demands* $\hat{\delta}_v^{\text{unf},+}$ for each station $v \in V$. They occur whenever the expected cumulated demand $\mu_v(t)$ over time horizon $t \in [0, \hat{t}_{\max}]$ cannot be satisfied, i.e., when $\mu_v(t) < 0 \wedge a_v(t) = 0$ or $\mu_v(t) > 0 \wedge a_v(t) = C_v$, respectively. Unfulfilled demands occurring at station v between events $j - 1$ and j, $j = 1, \ldots, |W_v|$, can formally be described as

$$\delta_{v,j}^{\text{unf},-} = \max(\mu_v(t_{l_j}^{i_j}) - \mu_v(t_{l_{j-1}}^{i_j-1}) - a_{v,j-1} + y_{l_j}^{i_j}, 0) \quad (4)$$

$$\delta_{v,j}^{\text{unf},+} = \max(-(\mu_v(t_{l_j}^{i_j}) - \mu_v(t_{l_{j-1}}^{i_j-1})) - (C_v - a_{v,j-1}) - y_{l_j}^{i_j}, 0), \quad (5)$$

and consequently the overall unfulfilled demands are

$$\hat{\delta}_v^{\text{unf},-} = \sum_{j=1}^{|W_v|} \delta_{v,j}^{\text{unf},-}, \quad \text{and} \quad \hat{\delta}_v^{\text{unf},+} = \sum_{j=1}^{|W_v|} \delta_{v,j}^{\text{unf},+}. \quad (6)$$

4.3 Classification of Stations

We assume that the stations are well-designed in a sense that their capacities are sufficiently large for daily fluctuations, i.e., it will not be necessary to pick up and deliver bikes to the same station at different times on a single day in order to fulfill all demands. Furthermore, we have shown in previous work [3] that the monotonicity restriction (i.e., it is allowed to either only pick up or deliver bikes at a station) has in practice only a neglectably small impact on the solution quality but substantially simplifies the problem. Additionally, our project partner Citybike Wien mentioned that they only perform either pickups or deliveries at a particular station on the same day. Therefore, we again classify the stations into pickup stations $V_{\text{pic}} \subseteq V$ and delivery stations $V_{\text{del}} \subseteq V$ and impose monotonicity, i.e., allow only the respective operations.

In the static case this classification is done by considering the total deviation in balance for a particular station, i.e., $p_v - q_v \ \forall v \in V$. If this value is less than 0, then the corresponding station refers to the set of delivery stations, and otherwise it is classified as a pickup station. However, in the dynamic case we have to consider user demands during the rebalancing process along with the scaling factors in the objective function (1).

Thus, we consider the situation when no rebalancing is done at all. Based on equation (6) and objective function (1) we determine for each station $v \in V$ the total penalty for slot deficit and bike deficit:

$$\delta_v^{\text{missing}} = \omega^{\text{unf}} \hat{\delta}^{\text{unf},+} + \omega^{\text{bal}} \min(0, a_{v,|W_v|} - q_v) - \\ \omega^{\text{unf}} \hat{\delta}^{\text{unf},-} - \omega^{\text{bal}} \min(0, q_v - a_{v,|W_v|}). \tag{7}$$

If $\delta_v^{\text{missing}} < 0$, v is a delivery station. If $\delta_v^{\text{missing}} > 0$, v is a pickup station. Otherwise, if $\delta_v^{\text{missing}} = 0$, the station is already balanced, and thus, will not be considered anymore.

4.4 Restrictions on Loading Instructions

For every stop of a vehicle, we need to calculate how many bikes the vehicle is allowed to pick up or deliver at most so that the capacity constraints are never violated and unnecessary unfulfilled demands are never introduced. These bounds are then utilized to set loading instructions for the corresponding vehicle stops later during the optimization process. Formally, we define slacks $\Delta y_{v,j}^-$ and $\Delta y_{v,j}^+$ as the maximum amount of bikes which may be delivered/picked up at station v and event $j = 1, \ldots, |W_v|$.

$$\Delta y_{v,j}^- = \begin{cases} \max(0, q_v - a_{v,|W_v|}) & \text{for } j = |W_v| \\ \min(C_v - a_{v,j}, \Delta y_{v,y+1} + \delta_{v,j+1}^{\text{unf},-}) & \text{for } j = 1, \ldots, |W_v| - 1 \end{cases} \tag{8}$$

$$\Delta y_{v,j}^+ = \begin{cases} \max(0, a_{v,|W_v|} - q_v) & \text{for } j = |W_v| \\ \min(a_{v,j}, \Delta y_{v,j+1} + \delta_{v,j+1}^{\text{unf},+}) & \text{for } j = 1, \ldots, |W_v| - 1. \end{cases} \tag{9}$$

Note, that we have to iterate backwards by starting with the last event until we reach the time when the currently considered vehicle stop occurs.

By definition, let $\Delta y_{v,j}^{\text{unf},+}$ and $\Delta y_{v,j}^{\text{unf},-}$ denote the slack without including the last event, i.e., starting with event $j = |W_v| - 1$. These two terms are used by the construction heuristic in the next section.

5 Greedy Construction Heuristic

We extend the Greedy Construction Heuristic (GCH) from our previous work [3] to fit the dynamic case. A vehicle tour is built by iteratively appending stations from a set of *feasible successors* $F \subseteq V$. This set includes each station which can be *improved* and is reachable within the vehicles time budget. An improvement may be achieved if $\delta_v^{\text{missing}} > 0$, $\forall v \in V_{\text{pic}}$, or $\delta_v^{\text{missing}} < 0$, $\forall v \in V_{\text{del}}$. Then, for each station $v \in F$ we compute the total number of bikes that can either be picked up from or delivered to this station:

$$\gamma_v = \begin{cases} \min(\Delta y_{v,j}^-, Z_l - b_l) & \text{for } v \in F \cap V_{\text{pic}}, \\ \min(\Delta y_{v,j}^+, b_l) & \text{for } v \in F \cap V_{\text{del}}. \end{cases} \tag{10}$$

Note that b_l denotes the number of bikes currently stored in vehicle l. As shown in (10), we need the slacks for determining possible loading instructions as they are calculated by equation (8). In order to guarantee that vehicles return empty to the depot, we correct the load for pickup stations by estimating the amount of bikes that can be delivered afterwards in the same fashion as in [3] by recursively looking forward.

It is necessary to consider impacts of loading instructions on a station with respect to target fill level and unfulfilled demands separately and weight them with ω_{bal} and ω_{unf} in the same way as it is done in the objective function (1). We obtain

$$g(v) = \begin{cases} \dfrac{\omega_{\mathrm{bal}} \cdot \min(\gamma_v, \max(0, a_{v,|W_v|} - q_v)) + \omega_{\mathrm{unf}} \cdot \min(\gamma_v, \Delta y_{v,j}^{\mathrm{unf},+})}{t_{u,v}} & \forall v \in V_{\mathrm{pic}}, \\[3mm] \dfrac{\omega_{\mathrm{bal}} \cdot \min(\gamma_v, \max(0, q_v - a_{v,|W_v|})) + \omega_{\mathrm{unf}} \cdot \min(\gamma_v, \Delta y_{v,j}^{\mathrm{unf},-})}{t_{u,v}} & \forall v \in V_{\mathrm{del}}, \end{cases} \tag{11}$$

where $t_{u,v}$ is the travel time from the vehicle's last stop u to station v. In each greedy iteration the station with the highest $g(v)$ is appended to the currently considered vehicle tour. Loading instructions are set as follows:

$$y_{v,j} = \gamma_v \text{ if } v \in V_{\mathrm{pic}}, \quad \text{and} \quad y_{v,j} = -\gamma_v \text{ if } v \in V_{\mathrm{del}} \tag{12}$$

Since in the dynamic case timing is important, we additionally introduce a term which we refer to as *urgency*. It states how urgent it is to visit stations with future unfulfilled demands. We propose two methods to compute this value.

Additive Urgency: For a station v we consider the time of the next period where unfulfilled demands occur. If the vehicle cannot reach the station until the first period starts, we consider the next period, and so on. In case a station has no periods of unfulfilled demands at all or none of them are reachable in time, it is ignored. Moreover, we introduce an additional scaling factor ω_{urg} which denotes the importance of urgency. Formally,

$$u_{\mathrm{add}} = \begin{cases} 0 & \text{if } t_v^{\mathrm{unf}} < t_{u,v} \\[2mm] \omega_{\mathrm{urg}} \cdot \dfrac{\delta_{\mathrm{unf}}}{t_v^{\mathrm{unf}}} & \text{if } t_v^{\mathrm{unf}} \geq t_{u,v} \end{cases} \tag{13}$$

where t_v^{unf} is the time left up to the start of the next unfulfilled demand for station $v \in V$ and $t_{u,v}$ is the travel time to the considered station which, by definition, has to be greater than 0. The greedy value including the urgency of the visit $g'(v)$ is then calculated as $g'(v) = g(v) + u_{\mathrm{add}}$.

Multiplicative Urgency: In the multiplicative approach we multiply the basic $g(v)$ from (11) by an *exponential function*. Again, we consider the time until the next unfulfilled demand, analogously as for the additive approach. The term is computed as

$$u_{\mathrm{mul}} = \exp(-\max(0, t_v^{\mathrm{unf}} - t_{u,v}) \cdot \omega_{\mathrm{urg}}). \tag{14}$$

The greedy value criterion is then extended to $g'(v) = g(v) \cdot u_{\mathrm{mul}}$.

PILOT Construction Heuristic: Due to the nature of greedy algorithms, shortsighted decisions cannot be completely avoided no matter how we choose the greedy evaluation criterion. Therefore, we use the PILOT method [18] to address this drawback. The functionality remains the same as in [5] which extends GCH by evaluating each potential successor in a deeper way by constructing a complete temporary route from it, and finally considering its objective value as $g(v)$.

6 Metaheuristic Approaches

In order to further improve the results obtained by the construction heuristics, we apply Greedy Randomized Adaptive Search (GRASP) and Variable Neighborhood Search (VNS). For both metaheuristic approaches we use an incomplete solution representation based on storing for each vehicle $l \in L$ its route $r_l = (r_l^1, \ldots, r_l^{\rho_l})$ only. The loading instructions $y_{l,v}^i$, $l \in L$, $v \in V$, $i = 1, \ldots, \rho_l$ are efficiently calculated during evaluation by applying the same greedy strategy as in GCH, see Section 5, utilizing the restriction procedure from Section 4.4 to obtain bounds on the y-variables and accelerate the calculations.

Variable Neighborhood Search: The VNS approach from [3] is adapted with respect to the procedure for deriving loading instructions. The general layout and neighborhood structures remain the same. We use remove-station, insert-unbalanced-station, intra-route-2-opt, replace-station, intra-or-opt, 2-opt*-inter-route-exchange and intra-route 3-opt neighborhood structures for local improvement within an embedded Variable Neighborhood Descent (VND), while for shaking we apply move-sequence, exchange-sequence, destroy-&-recreate, and remove-stations operations.

Greedy Randomized Adaptive Search: We also consider GRASP by extending the construction heuristics in the same way as in our previous work [5] with adaptations for the dynamic problem variant. The idea is to iteratively apply GCH or PILOT from Section 5 and locally improve each solution with VND. While there we always select the best successor, we use for GRASP a restricted candidate list with respect to the greedy evaluation criterion. The degree of randomization is controlled by a parameter $\alpha \in [0, 1]$. In the dynamic case we used the same values which turned out to work best in the static case.

7 Computational Results

We performed comprehensive tests for our DBBSS approaches. Generating new benchmark instances was necessary in order to introduce the user demand values. They are based on the same set of Vienna's real Citybike stations we used in our previous works [2–5]. Cumulated user demands for the stations are piecewise linear functions derived from historical data based on an hourly discretization. The instances we use in this paper contain 30 to 90 stations with different numbers of vehicles and different time budgets and are available at[1]. As the BSS in

---

[1] https://www.ads.tuwien.ac.at/w/Research/Problem_Instances#bbss

Table 1. Results of GCH, PILOT, and the variants with VND

Inst. set			GCH				PILOT				GCH-VND				PILOT-VND			
$|V|$	$|L|$	$\hat{t}_{max}$	#best	$\overline{obj}$	sd	$\widetilde{t_{tot}}$ [s]	#best	$\overline{obj}$	sd	$\widetilde{t_{tot}}$ [s]	#best	$\overline{obj}$	sd	$\widetilde{t_{tot}}$ [s]	#best	$\overline{obj}$	sd	$\widetilde{t_{tot}}$ [s]
30	1	8h	0	54.06	12.50	< 0.1	0	50.98	11.19	0.1	18	50.61	11.56	0.4	13	49.97	10.95	0.4
30	2	4h	0	59.79	15.65	< 0.1	1	55.47	13.78	< 0.1	9	55.87	13.58	0.2	22	54.89	13.44	0.1
60	1	8h	0	186.49	28.14	< 0.1	1	180.59	28.81	0.5	8	180.20	28.72	0.5	23	178.85	29.29	0.9
60	2	4h	0	202.69	31.82	< 0.1	0	191.06	29.98	0.2	9	193.32	30.06	0.4	21	189.69	29.81	0.4
60	2	8h	0	104.49	12.77	< 0.1	0	98.64	11.03	0.9	12	98.00	12.05	2.4	18	96.74	10.80	3.2
60	4	4h	0	118.76	17.38	< 0.1	0	106.98	13.53	0.4	4	108.96	13.34	1.8	26	105.36	13.41	1.3
90	1	8h	0	354.83	34.79	< 0.1	0	346.73	33.49	1.1	6	348.20	34.74	0.7	24	344.99	33.45	1.5
90	2	4h	0	371.13	34.55	< 0.1	0	360.49	36.06	0.5	5	362.74	35.53	0.5	25	358.21	35.09	0.9
90	2	8h	0	232.86	27.07	< 0.1	0	221.02	24.24	2.1	3	223.16	24.71	2.6	27	218.57	23.86	4.1
90	3	8h	0	155.26	19.27	< 0.1	0	144.35	16.80	2.8	6	144.43	17.94	8.6	24	141.25	16.26	6.9
90	4	4h	0	254.25	27.51	< 0.1	0	239.70	27.86	1.0	7	242.91	27.90	2.1	23	237.47	27.63	2.2
90	5	4h	0	210.12	24.26	< 0.1	0	194.03	24.55	1.2	7	195.75	24.38	4.2	23	191.72	23.96	3.3
		Total	0	2304.73	285.72	< 0.1	2	2190.04	271.31	10.8	94	2204.15	274.48	24.4	269	2167.71	267.95	25.2

Table 2. Results of static VNS, dynamic VNS, and PILOT-GRASP

Inst. set			SVNS			DVNS			PILOT-GRASP		
$|V|$	$|L|$	$\hat{t}_{max}$	#best	$\overline{obj}$	sd	#best	$\overline{obj}$	sd	#best	$\overline{obj}$	sd
30	1	8h	4	54.90	10.93	20	47.36	10.51	9	47.54	10.57
30	2	4h	4	58.68	13.08	19	50.84	11.50	11	50.84	11.35
60	1	8h	0	197.25	30.01	29	172.30	27.13	4	172.84	26.96
60	2	4h	0	207.09	30.37	22	182.12	29.28	10	183.09	29.13
60	2	8h	0	114.64	12.75	26	91.80	10.63	4	92.30	10.40
60	4	4h	0	126.42	15.11	23	99.24	11.39	7	99.85	11.54
90	1	8h	0	368.18	37.47	19	337.92	32.74	11	338.49	32.23
90	2	4h	0	380.50	38.80	19	349.98	33.97	11	351.05	34.74
90	2	8h	0	242.60	26.09	11	210.62	23.79	19	210.19	23.03
90	3	8h	0	168.99	16.31	11	135.97	15.10	19	135.40	15.06
90	4	4h	0	262.41	30.41	17	225.94	25.77	13	226.39	26.19
90	5	4h	0	216.53	23.76	18	182.03	21.82	12	182.20	22.21
		Total	8	2398.19	285.10	234	2086.11	253.63	130	2090.16	253.38

Vienna currently consists of 111 stations and 1300 bicycles the instances used inhere are realistic and praxis-relevant. For each parameter combination exists a set of 30 independent instances. All our algorithms are implemented in C++ using GCC 4.6. Each test run was performed on a single core of an Intel Xeon E5540 machine with 2.53 GHz. The scaling factors of the objective function are set to $\omega^{unf} = \omega^{bal} = 1$, $\omega^{load} = \omega^{time} = \frac{1}{100\,000}$, i.e., an improvement with respect to balance and/or unfulfilled demands is always preferred over reducing the tour length and/or the number of loading instructions. For the greedy evaluation criterion we use multiplicative urgency. We omit a detailed comparison since the difference between these strategies becomes more significant only on larger instances with hundreds of stations.

In Table 1 we compare different methods for quickly obtaining good starting solutions, namely GCH, PILOT, GCH with VND, and PILOT with VND. The columns show the instance characteristics, and for each algorithm the number of times the corresponding approach yields the best result (#best), the average objective values ($\overline{obj}$), the standard deviations (sd), and the average CPU-times $\widetilde{t_{tot}}$. The differences of the average objective values are frequently relatively small due to the weight factors ω^{load} and ω^{time}, but they are still crucial for evaluating the quality of solutions. Therefore, the #best numbers give us a better indication of which algorithm variants perform best. We observe that PILOT outperforms GCH on every instance while the additional time is only moderate. This trend continues when we add VND to further improve the solutions. Not only does PILOT-VND outperform GCH-VND, but it also requires less time. This is due to the superior starting solutions, so VND terminates after fewer iterations.

In Table 2 we test our metaheuristic approaches and additionally compare them to the VNS for the static case from [3], denoted by SVNS. For a reasonable comparison, SVNS initially converts a DBBSS instance into a static one by adding for each station the final cumulative user demand to the respective target value; negative values and values exceeding the station capacity C_v are replaced by zero and C_v, respectively. The idea is to neglect the timing aspects of station visits and check if this static VNS is able to find reasonable solutions also for the dynamic case. To assure always obtaining feasible solutions to DBBSS in the end, loading instructions for the finally best static solution are recalculated by the new greedy strategy of the dynamic case. We observe that GCH from Table 1 already performs a little bit better than the SVNS. DVNS and GRASP are able to compute results that are more than 10% better than those of SVNS. Therefore, we conclude that although it is possible to apply algorithms for the static case to the dynamic scenario, dedicated dynamic approaches taking time-dependent user demands into account are clearly superior. Among the dynamic approaches DVNS performs best on most of the considered instances. According to a Wilcoxon signed-rank test, all observed differences on the overall number of best solutions among any pair of compared approaches are statistically significant with an error level of less than 1%.

8   Conclusions and Future Work

In this work we showed how to extend the metaheuristics developed in our previous work for static BBSS to the significantly more complex dynamic variant. Starting from a model which can handle essentially arbitrary time-dependent expected user demand functions, we proposed an efficient way to calculate loading instructions for vehicle tours. We use an objective function where the weights of unfulfilled user demands and target fill levels can be adjusted in an easy way. Practically, this has a high relevance for the BSS operator. We also extended our previously introduced construction heuristics, VNS and GRASP, so that dynamic user demands are considered appropriately. Tests on practically realistic instances show that the dynamic approaches indeed make sense. Depending on the available time for optimization, greedy or PILOT construction heuristics are useful for fast runs, while VNS is most powerful for longer runs.

In future work it would be particularly interesting to also consider the impact of demand shifts among stations when their neighbors become either full or empty. Especially, when users want to return bikes and an intended target station is full, this demand obviously will not diminish but be shifted to some neighboring station(s). Considering this aspect might lead to an even more precise model, but also increases the model's complexity significantly.

References

1. DeMaio, P.: Bike-sharing: History, impacts, models of provision, and future. Journal of Public Transportation 12(4), 41–56 (2009)
2. Raidl, G.R., Hu, B., Rainer-Harbach, M., Papazek, P.: Balancing bicycle sharing systems: Improving a VNS by efficiently determining optimal loading operations. In: Blesa, M.J., Blum, C., Festa, P., Roli, A., Sampels, M. (eds.) HM 2013. LNCS, vol. 7919, pp. 130–143. Springer, Heidelberg (2013)
3. Rainer-Harbach, M., Papazek, P., Hu, B., Raidl, G.R.: Balancing bicycle sharing systems: A variable neighborhood search approach. In: Middendorf, M., Blum, C. (eds.) EvoCOP 2013. LNCS, vol. 7832, pp. 121–132. Springer, Heidelberg (2013)
4. Rainer-Harbach, M., Papazek, P., Hu, B., Raidl, G.R.: PILOT, GRASP, and VNS approaches for the static balancing of bicycle sharing systems. Technical Report TR 186-1-13-01, Vienna, Austria (29 pages, 2013, submitted to the JOGO)
5. Papazek, P., Raidl, G.R., Rainer-Harbach, M., Hu, B.: A PILOT/VND/GRASP hybrid for the static balancing of public bicycle sharing systems. In: Moreno-Díaz, R., Pichler, F., Quesada-Arencibia, A. (eds.) EUROCAST. LNCS, vol. 8111, pp. 372–379. Springer, Heidelberg (2013)
6. Chemla, D., Meunier, F., Calvo, R.W.: Bike sharing systems: Solving the static rebalancing problem. Discrete Optimization 10(2), 120–146 (2013)
7. Benchimol, M., Benchimol, P., Chappert, B., De la Taille, A., Laroche, F., Meunier, F., Robinet, L.: Balancing the stations of a self service bike hire system. RAIRO – Operations Research 45(1), 37–61 (2011)
8. Raviv, T., Tzur, M., Forma, I.A.: Static repositioning in a bike-sharing system: models and solution approaches. EURO Journal on Transp. and Log., 1–43 (2013)
9. Di Gaspero, L., Rendl, A., Urli, T.: A hybrid ACO+CP for balancing bicycle sharing systems. In: Blesa, M.J., Blum, C., Festa, P., Roli, A., Sampels, M. (eds.) HM 2013. LNCS, vol. 7919, pp. 198–212. Springer, Heidelberg (2013)
10. Di Gaspero, L., Rendl, A., Urli, T.: Constraint-based approaches for balancing bike sharing systems. In: Schulte, C. (ed.) CP 2013. LNCS, vol. 8124, pp. 758–773. Springer, Heidelberg (2013)
11. Contardo, C., Morency, C., Rousseau, L.M.: Balancing a dynamic public bike-sharing system. Technical Report CIRRELT-2012-09, Montreal, Canada (2012)
12. Schuijbroek, J., Hampshire, R., van Hoeve, W.J.: Inventory Rebalancing and Vehicle Routing in Bike Sharing Systems. Technical Report 2013-E1, Tepper School of Business, Carnegie Mellon University (2013)
13. Chemla, D., Meunier, F., Pradeau, T., Calvo, R.W., Yahiaoui, H.: Self-service bike sharing systems: simulation, repositioning, pricing. Technical Report hal-00824078, CERMICS (2013)
14. Pfrommer, J., Warrington, J., Schildbach, G., Morari, M.: Dynamic vehicle redistribution and online price incentives in shared mobility systems. Technical report, Cornell University, NY (2013)
15. Lin, J.H., Chou, T.C.: A geo-aware and VRP-based public bicycle redistribution system. International Journal of Vehicular Technology (2012)
16. Lin, J.R., Yang, T.H., Chang, Y.C.: A hub location inventory model for bicycle sharing system design: Formulation and solution. Computers & Industrial Engineering 65(1), 77–86 (2013)
17. Nair, R., Miller-Hooks, E., Hampshire, R.C., Bušić, A.: Large-scale vehicle sharing systems: Analysis of Vélib'. Int. Journal of Sustain. Transp. 7(1), 85–106 (2013)
18. Voß, S., Fink, A., Duin, C.: Looking ahead with the PILOT method. Annals of Operations Research 136, 285–302 (2005)

Cooperative Selection:
Improving Tournament Selection via Altruism

Juan Luis Jiménez Laredo[1], Sune S. Nielsen[1], Grégoire Danoy[1],
Pascal Bouvry[1], and Carlos M. Fernandes[2]

[1] FSTC-CSC/SnT,
University of Luxembourg, 6, rue Richard Coudenhove-Kalergi,
1359 Luxembourg, Luxembourg
{juan.jimenez,sune.nielsen,gregoire.danoy,pascal.bouvry}@uni.lu
[2] Laseeb: Evolutionary Systems and Biomedical Engineering,
Technical University of Lisbon, Lisbon, Portugal
cfernandes@laseeb.org

Abstract. This paper analyzes the dynamics of a new selection scheme based on altruistic cooperation between individuals. The scheme, which we refer to as cooperative selection, extends from tournament selection and imposes a stringent restriction on the mating chances of an individual during its lifespan: winning a tournament entails a depreciation of its fitness value. We show that altruism minimizes the loss of genetic diversity while increasing the selection frequency of the fittest individuals. An additional contribution of this paper is the formulation of a new combinatorial problem for maximizing the similarity of proteins based on their secondary structure. We conduct experiments on this problem in order to validate cooperative selection. The new selection scheme outperforms tournament selection for any setting of the parameters and is the best trade-off, maximizing genetic diversity and minimizing computational efforts.

1   Introduction

Evolutionary algorithms (EAs) are optimization meta-heuristics inspired by the Darwinian process of natural selection. As in nature, individuals in EAs compete for survival and the fittest are preferentially selected for reproduction. Through the course of generations, evolution operates bottom up by filtering good genes in the population and optimizing the design of the individuals. This complex process is triggered by a set of simple rules, the cornerstones being the breeding operators and the selection scheme: the formers leveraging the proper mixing of individuals' structures and the latter balancing the selection pressure. If the selection pressure is too intensive, the population will lose genetic diversity quickly and the algorithm will converge to local optima. On the other hand, if the selection pressure is too relaxed, the speed of convergence will slow down and the algorithm will waste computational efforts. Hence, the design of efficient selection schemes remains as an open topic of research in Evolutionary Computation.

C. Blum and G. Ochoa (Eds.): EvoCOP 2014, LNCS 8600, pp. 85–96, 2014.

In addition to canonical approaches such as ranking or roulette wheel [4], other selection schemes have been designed, e.g. to balance exploration and exploitation [1], to autonomously decide on the state of reproduction of individuals [7], or to be able to self-adapt the selection pressure on-line [3]. Nevertheless, tournament selection [8] is still one of the most studied and employed schemes for selection. Tournament selection is an easy-to-implement easy-to-model scheme, in which the best of t randomly sampled individuals is selected for mating. This simple scheme has many advantages such as allowing the tuning of the selection pressure by simply varying the size of the tournament, having a low algorithmic complexity (i.e. $O(t)$ for a single selection) and being suitable for parallel and decentralized execution as it does not require to compute estimates of the global population. However, tournament selection has been also reported to have some disadvantages that we aim to cover in this paper.

Poli [9] points out two different phases in a tournament: the *sampling* phase where t individuals are randomly chosen from the population, and the *selection* phase, in which the best of the sampled individuals is selected for reproduction. The two main issues arising from Poli's study are known as the *not-sampling* and *multi-selection* problems, both of them responsible for the loss of diversity. The *not-sampling* issue refers to the probability of an individual for not being sampled. Xie, Zhang and Andreae [14] conduct a thorough analysis on this issue and conclude that, independently of the population size, some individuals in every generation will never be sampled when using small tournament sizes, e.g. for binary tournament 13% of the individuals are never sampled, if $t = 3$ that value is $\sim 5\%$, and so on. In contrast, the *multi-selection*[1] issue refers to the probability of an individual for being selected multiple times in the selection phase. The risk here is that, if the tournament size is large, a small elite of individuals could take over the population and quickly reduce the diversity. Given that the only tunable parameter in tournament selection is the tournament size, either the *not-sampling* or *multi-selection* issue will always be present to some extent at any value of the parameter.

Some authors have presented alternative solutions to the *not-sampling* problem. The backward-chaining proposal by Poli [9], the unbiased tournament selection by Sokolov and Whitley [12] or the round-replacement by Xie and Zhang [13] successfully address the issue. In their respective proposals, all authors are able to save the computational efforts associated to the evaluation of not-sampled individuals. Xie and Zhang, however, show that the sampling countermeasures are only effective in the case of binary tournament. For larger tournament sizes they claim that *"the not sampled issue in standard tournament selection is not critical"*. On the basis of previous findings the authors recommend that research should focus on tuning the parent selection pressure instead of developing alternative sampling replacement strategies.

---

[1] The *multi-selection* problem is alternatively referred to as *not-selection* problem, i.e. the multiple selection of some individuals implies the opposite problem, some individuals are never selected.

With such a recommendation in mind, cooperative selection [6] tackles the *selection* phase in tournament selection. The aim is to minimize the loss of diversity due to the *multi-selection* of elite individuals. Given that the *multi-selection* issue is more relevant for large tournament sizes, the *not-sampling* of individuals is assumed as a minor issue. Nonetheless, any of the aforementioned *not-sampling* solutions could be easily integrated within our approach as they are complementary. Cooperative selection tries to prevent the *multi-selection* of elite individuals by implementing an altruistic rule: after winning a tournament, an individual must decrease its own fitness as to yield its position to the second-ranked individual for future tournaments.

In this paper, we analyze first the properties of the approach under simplified assumptions. Then, in order to validate such properties in a real problem, we tackle a new combinatorial problem designed to maximize the similarity between proteins. In that context, cooperative selection is shown to preserve a higher genetic diversity than tournament selection and to require less evaluations to yield the same quality in solutions.

The rest of the paper is structured as follows. Section 2 introduces the new selection scheme. This section also analyzes the main properties of the approach and provides some assessment for tuning the scheme parameters. Section 3 presents a new combinatorial problem that is used afterwards to validate the cooperative selection scheme. Finally, some conclusions and future lines of research are exposed in section 4.

2   Cooperative Selection

Cooperative selection is a variant in the *selection phase* of tournament selection. This section presents a description of the new selection scheme and analyzes its properties for different parameter values using tournament selection as the counterpart for comparison.

2.1   Scheme Description

In order to describe cooperative selection, let us first introduce the following definitions and nomenclature:

- An *association* $\vec{a} = \{\vec{x_1}, \ldots, \vec{x_a}\}$ is the scheme's equivalent of a *tournament*, where a set of individuals $\vec{x_i} : i \in \{1, \ldots, a\}$ is randomly sampled from a population P to compete for selection in an association of size a.
- Every individual $\vec{x} = \{x_1, \ldots, x_l\} \mid \vec{x} \in P$ has a length of l parameters and is characterized by two different fitness metrics:
  - The canonical fitness function $f(\vec{x})$.
  - A transcription of $f(\vec{x})$ called cooperative fitness $f_{coop}(\vec{x})$, which is initialized in an atomic operation:

$$sync.eval(\vec{x})\{f(\vec{x}), f_{coop}(\vec{x}) \leftarrow f(\vec{x})\} \tag{1}$$

Hence, $f_{coop}(\vec{x}) = f(\vec{x})$ holds true right after the evaluation of $\vec{x}$.

Given an association $\overrightarrow{a}$ as provided by the sampling method described above, Procedure 1 details the criteria to select an individual from $\overrightarrow{a}$. Note here that $\overrightarrow{a}$ and an analogue tournament $\overrightarrow{t}$ are indistinguishable and therefore, the only difference between cooperative and tournament selection relies on this procedure.

Procedure 1. Pseudo-code of Cooperative Selection

CooperativeSelection ($\overrightarrow{a}$)

#1. Ranking step:
$$rank(\overrightarrow{a}) = \{(\overrightarrow{x}_1, \ldots, \overrightarrow{x}_i, \overrightarrow{x}_{i+1}, \ldots, \overrightarrow{x}_a) \mid \forall i : f_{coop}(\overrightarrow{x}_i) \geq f_{coop}(\overrightarrow{x}_{i+1})\}$$

#2. Competition step:
$$winner \leftarrow \overrightarrow{x}_1$$

#3. Altruistic step:
$$f_{coop}(\overrightarrow{x}_1) \leftarrow \frac{f_{coop}(\overrightarrow{x}_2) + f_{coop}(\overrightarrow{x}_3)}{2}$$

return *winner*

The competition (or *selection* phase) in cooperative selection consists of three steps. In the first one, the sampled individuals ($\overrightarrow{a}$) are ranked according to their cooperative fitness. To that end, without any loss of generality, we have considered in this paper the case of maximizing the fitness. In a second step, the individual with the highest cooperative fitness $f_{coop}(\overrightarrow{x}_1)$ is selected as the winner of the competition. However, these two steps would not change from a canonical competition if it were not for the final step where, after being selected, the *winner* altruistically decreases its own cooperative fitness. In the next section we show that such a simplistic rule has a tremendous impact on reducing the environmental selection pressure and consequently the loss of genetic diversity.

2.2   Properties and Tuning of Parameters

Cooperative selection is based on tournament selection and, as in tournament selection, the scheme can also be divided into two main phases: the *sampling* phase in which the individuals are randomly sampled to create an association, and the *selection* phase in which the individual with the highest cooperative fitness is selected for reproduction. In order to analyze the properties of cooperative selection, this section tackles both phases incrementally by assessing first the bias introduced during the *sampling* phase, and then by analyzing the combined effect of the *not-sampling* and *not-selection* of individuals during the *selection* phase.

The Sampling Phase. The *sampling* of individuals in cooperative selection can be described by the probability model presented by Xie et al. [14] for tournament selection. The model states that the probability of an individual to be sampled in $y \in \{1, \ldots, P\}$ associations/tournaments is:

$$1 - \left(1 - \frac{1}{P}\right)^{ya} \qquad (2)$$

where P is the population size, a the association/tournament size, and y the number of associations/tournaments within the generation.

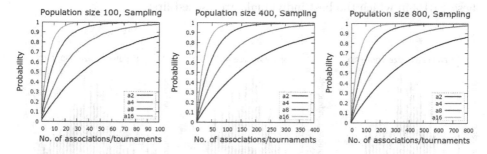

Fig. 1. Probability of sampling an individual at least once per generation for $y = P$. Given that the sampling of individuals is performed identically in tournament and cooperative selection, these results hold in both schemes. Results are the average of 30 independent runs.

In Fig. 1 we verify empirically the model via simulations and estimate the sampling probabilities for $P \in \{100, 400, 800\}$ and $a \in \{2, 4, 8, 16\}$. The obtained results show that the association size (a) is the main factor influencing the sampling probability while the population size (P) does not seem to affect the probability trend (Sokolov and Whitley [12] show that the population size may indeed affect the trend if $P < 10$ however populations in EAs are usually not that small). Independently of the population size, there is a certain probability for an individual of not being sampled when a is small. An individual that is not sampled does not have any chance to pass to subsequent generations and thus, its associated genetic material and computational efforts are hopelessly wasted. Such a waste amounts to $\sim 13\%$ of the population for $a = 2$; a percentage that becomes progressively smaller as the association size increases and that can be neglected from $a > 5$. Therefore, the association size should be preferentially set to large values for an optimal operation of the scheme.

The Selection Phase. Once an association $\overrightarrow{a} = \{\overrightarrow{x_1}, \ldots, \overrightarrow{x_a}\}$ is sampled from the population, the *selection* phase is responsible for determining which individual in $\overrightarrow{a}$ will be finally selected for reproduction. At this stage, a large association size entails an increase in the selection pressure and a loss of genetic diversity. Such an effect is well known in tournament selection for large tournament sizes and needs to be assessed for cooperative selection.

Xie and Zhang [13] propose an experimental method for assessing the loss of diversity according to different values of a. The method employs a synthetic population called *fitness rank distribution* (FRD). The idea behind a FRD is that a population can be partitioned into different bags of individuals with an equal fitness. Given that each of these bags has an associated fitness rank, a selection scheme can be characterized by the amount of individuals which are selected from the different bags. In order to analyze cooperative selection, we will assume a simplistic scenario in which the FRDs emulate the fitnesses of a randomly initialized population in an easy problem [11]. Fig. 2 shows three of these FRDs in which the best individuals are ranked first.

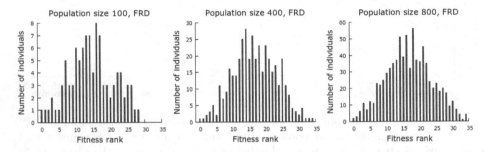

Fig. 2. Fitness rank distributions for three different population sizes. FRDs are dropped from a normally distributed variable $X \sim \mathcal{N}(0, 36)$ and then shifted until $min(X) = 0$. This method approximates a $\mathcal{N}(18, 36)$ distribution and guarantees non-negative values.

Using a FRD as an input, our genetic algorithm assumes a generational scheme with selection as the single operator to be applied. Any possible loss of diversity is therefore due to the selection scheme and can be assessed by computing the amount of individuals discarded from one generation to the other. In this context, there are two possible causes for losing diversity: those individuals that are *not-sampled*, and those ones that being sampled are *not-selected*.

Fig. 3 compares the loss of diversity in tournament and cooperative selection as a function of the tournament/association size. The selection pressure monotonically increases for both schemes. However, cooperative selection scales more gracefully than tournament selection: for $a = t = 20$ cooperative selection outperforms tournament selection by a $\sim 24\%$.

Given the different responses of both schemes to parametrization, we define as *analogues* a tournament (t) and an association (a) inducing the same loss of diversity, e.g. $t = 2$ and $a = 8$ in Fig. 3. This notion is relevant to make comparisons between schemes: since two analogue parameters preserve the diversity equally, one scheme will outperform the other if, for an analogue parametrization, the scheme preferentially selects fitter individuals. Note here that the variety and quality of the selected individuals are two conflicting objectives.

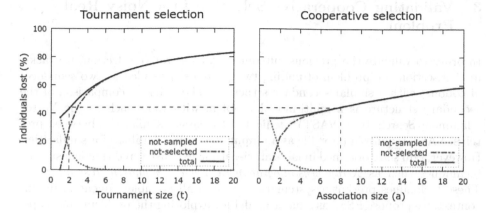

Fig. 3. Loss of diversity in tournament selection (*left*) and cooperative selection (*right*) for $P = 400$. The dashed line exemplifies an equivalence between analogue tournament and association sizes.

To gain insights into this question, Fig. 4 computes the selection frequencies for different parametrizations of tournament and cooperative selection. The results show two main outcomes:

1. Cooperative selection is more robust to parameter tuning than tournament selection as the size of an association has less impact on the selection frequencies than the respective tournament size.
2. In the case of an analogue parametrization (i.e. $t2$ is analogue to $a8$), the frequency of good individuals is higher in cooperative selection than in tournament selection.

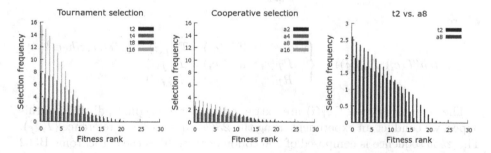

Fig. 4. Expected selection frequencies for different tournament and association sizes in tournament selection (*left*) and cooperative selection (*center*). The graph on the right compares the selection frequencies between binary tournament ($t2$) and its analogue association size ($a8$). Frequencies are obtained from 30 independent runs.

3   Validating Cooperative Selection in a Noisy Real Problem

In order to validate the previous outcomes on cooperative selection, we tackle in this section the problem of finding two proteins (provided as two sequences of codons) with a similar secondary structure. The pairwise comparison of the secondary structure is a multi-modal problem with tools such as the Vector Alignment Search Tool (VAST)[2] dedicated to classify similarities between proteins. VAST requires a pdb[3] file as an input to query a database for similarities. However, pdb's are obtained in laboratories using expensive and time-consuming techniques such as *X-ray crystallography* or *nuclear magnetic resonance* [2]. These techniques prevent a systematic search of all possible sequences. In this context, we propose ESSS as a meta-model for exploring the problem landscape.

3.1   Estimated Secondary Structure Similarity (ESSS)

The secondary structure is a primary approach to determine the three-dimensional form of an amino acid sequence. Given that the biological functionality of a protein is conferred by its 3D structure, establishing the 3D similarity between two proteins provides an estimation of their related properties. ESSS is an assessment on the similarity of the secondary structure between two sequences.

Using the tool *PROFphd* [10], the likely secondary structure type $T(c_i)$ can be estimated per codon c_i with a reliability $R_T(c_i) \in \{1...10\}$ in the sequence $C = \{c_1...c_N\}$. With $T_{ref}(i)$ the actual type found at position i of the reference structure, the estimated secondary structure similarity score F is calculated as follows:

$$F(C, T_{ref}) = \sum_{i=1}^{N} Match(T(c_i), T_{ref}(i)) \tag{3}$$

where

$$Match(T(c_i), T_{ref}(i)) = \left\{ \begin{array}{ll} 0 & \text{if } T(c_i) \wedge T_{ref}(i) \notin \{Helix, Sheet\} \\ R_T(c_i) & \text{if } T(c_i) = T_{ref}(i) \\ -R_T(c_i) & \text{if } T(c_i) \neq T_{ref}(i) \end{array} \right.$$

The reference types $T_{ref}(i)$ are extracted from the original pdb file. In this paper, we conduct an experiment using the *256b* sequence[4] as a reference (T_{ref}). The *256b* sequence is composed of 106 codons and codifies the cytochrome B562 molecule.

---

[2] VAST and pdb's of sequences are available on-line at
http://www.ncbi.nlm.nih.gov/Structure/
[3] Protein Data Bank format.
[4] VAST and pdbs of sequences are available on-line at
http://www.ncbi.nlm.nih.gov/Structure/

Multi-modality. ESSS is a multi-modal problem. Only for the *256b* sequence that we tackle in this paper, VAST provides a list of 1635 related structures from which *4JEA* is the most similar. *4JEA* codifies the soluble cytochrome B562 and differs in only 13 out of 106 codons.

Noisiness. ESSS theoretical optimum is at $F(256b, 256b) = 860$ for the instance under study. However, to yield such a value, *PROFphd* should return the maximum reliability $R_T = 10$ at any of the 86 positions of the sequence folding as a helix. Given that *PROFphd* is a meta-model based on a neural network, there is a certain error due to the training process which translates in the actual fitness being $F(256b, 256b) = 463$. The noisiness of the function can be further proven since the sequence *4JEA* scores $F(4JEA, 256b) = 520$; more than the reference sequence itself, i.e. $F(4JEA, 256b) > F(256b, 256b)$.

3.2 Experimental Setup

Experiments in this paper are conducted for a generational 1-elitism genetic algorithm. Two versions of the algorithm are considered: one using cooperative selection and the other being tournament selection. In order to validate the properties described in section 2.2, *analogue* parameters $t = 2$ and $a = 8$ are tested. Additionally, $t = 8$ is also analyzed to check differences with respect to $a = 8$. Table 1 presents a summary of all parameter settings.

Table 1. Parameters of the experiments

ESSS instance	
Reference pdb file	256b
Individual length (L)	106
Features	*multimodality and noisiness*

GA settings	
Scheme	1-elitism generational GA
Selection of Parents	Tournament Selection & Cooperative Selection
Tournament size	2, 8
Association size	8
Recombination	Uniform Crossover, $p_c = 1.0$
Mutation	Uniform, $p_m = \frac{1}{L}$
Population sizes (P)	$\{100, 200, 400, 800\}$

3.3 Analysis of Results

Miller and Goldberg [8] were the first in analyzing the relations between noisy functions and selection schemes. In general, noise has a disruptive effect on the convergence of a genetic algorithm, delaying the convergence rate and increasing the computational requirements. The common approach to counteract this effect is to increase the selection intensity. However, if the selection pressure is too high, the algorithm risks to prematurely converge to sub-optimal solutions. Noisy functions are therefore an attractive context to validate the properties

of cooperative selection: the new selection scheme has been shown to increase
the selection frequency of the fittest individuals (i.e. selection intensity) without
paying an additional cost in diversity.

Unlike other noisy problems (see e.g. [5] for a survey), ESSS does not include
a true fitness function to recalibrate errors of the meta-model. Therefore, diversity is prioritized over quality in such a way that results can be validated *a
posteriori* in the lab [2]. That leads to the notion of *feasible region*: a range of
fitness values used as termination criteria. We assume the *feasible region* to be
$[F(256b, 256b), F(4JEA, 256b)] = [463, 520]$ for the instance under study.

Fig. 5 shows that cooperative selection ($a = 8$) yields better results than
tournament selection ($t = 2$) while the genetic diversity is similarly preserved in
both schemes as it was predicted in section 2.2 for *analogue* parametrizations. On
the other hand, tournament selection converges faster than cooperative selection
if the algorithms are equally parametrized ($a = 8 = t$). However, a high value
of tournament selection ($t = 8$) increases the selection pressure and the genetic
diversity is significantly diminished.

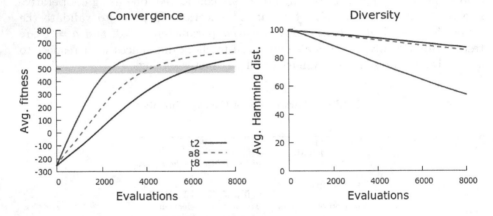

Fig. 5. Convergence of the average fitness in cooperative and tournament selection
(*left*) and respective average hamming distance (*right*). The area in gray denotes the
feasible region. Results are obtained over 30 independent runs for $P = 400$.

In order to find best trade-offs between speed of convergence and preservation
of diversity, we reproduce previous experiments for $P \in \{100, 200, 400, 800\}$ with
$\overline{F} = 520$ set as the termination condition. Fig. 6 shows the results for cooperative and tournament selection where the upper-left corner is the optimal area.
In all cases and independently of the population size, cooperative selection has
better results regarding diversity and number of evaluations. That is, cooperative selection outperforms tournament selection. Although such results must be
interpreted under the perspective of this study, the obtained conclusions should
be easily generalized to further problems and optimization paradigms.

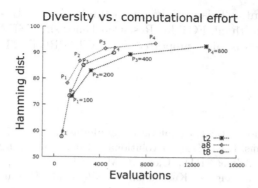

Fig. 6. Trade-off between diversity and computational effort for cooperative selection {$a8$} and tournament selection {$t2, t8$}

4   Conclusions and Future Works

This paper analyzes a new selection scheme that we call *cooperative selection* in the context of genetic algorithms. The scheme is an extension of tournament selection that, by implementing an altruistic behavior in winners of the competition, is able to cope with two of the main problems of tournament selection: the *"not-sampling"* and *"multi-selection"* of individuals. On the one hand, the approach preferentially uses large association sizes, which neglects the probability of *"not-sampling"* an individual. On the other hand, the depreciation of the fitness in winners of the competition diminishes the *"multi-selection"* of the same elite of individuals.

We show that the new scheme preserves a higher genetic diversity than tournament selection when both schemes are equally parametrized. Additionally, cooperative selection casts fitter individuals when the loss of diversity is equal in both schemes. That allows the algorithm to converge faster to quality solutions while preserving the genetic diversity. The key to explain such a behavior is that the new scheme can select the same variety of individuals as in tournament selection but choosing preferentially fitter ones.

In order to validate such properties in a real problem, we formulate in this paper a new combinatorial problem for maximizing the similarity between proteins. The conducted experiments confirm previous conclusions: cooperative selection outperforms tournament selection for any setting of the parameters and is the best trade-off, maximizing genetic diversity and minimizing computational efforts.

As a future work, we plan to proceed with the mathematical modeling of the proposed selection scheme. Additionally, we also plan to investigate the adequacy of the approach in high-dimensional continuous optimization problems.

Acknowledgements. This work has been funded by the UL-EvoPerf project. Fernandes wishes to thank FCT his Research Fellowship (SFRH/ BPD/66876/ 2009). This work was also supported by FCT PROJECT [PEst-OE/EEI/ LA0009/2013].

References

1. Alba, E., Dorronsoro, B.: The exploration/exploitation tradeoff in dynamic cellular genetic algorithms. IEEE Trans. Evolutionary Computation 9(2), 126–142 (2005)
2. Brünger, A.T., Adams, P.D., Clore, G.M., DeLano, W.L., Gros, P., Grosse-Kunstleve, R.W., Jiang, J., Kuszewski, J., Nilges, M., Pannu, N.S., Read, R.J., Rice, L.M., Simonson, T., Warren, G.L.: Crystallography and NMR System: A New Software Suite for Macromolecular Structure Determination. Acta Crystallographica Section D 54(5), 905–921 (1998)
3. Eiben, A.E., Schut, M.C., De Wilde, A.R.: Boosting genetic algorithms with self-adaptive selection. In: Proceedings of the IEEE Congress on Evolutionary Computation, pp. 1584–1589 (2006)
4. Eiben, A.E., Smith, J.E.: Introduction to Evolutionary Computing. Springer (2003)
5. Jin, Y., Branke, J.: Evolutionary optimization in uncertain environments-a survey. Trans. Evol. Comp. 9(3), 303–317 (2005)
6. Laredo, J.L.J., Dorronsoro, B., Fernandes, C., Merelo, J.J., Bouvry, P.: Oversized populations and cooperative selection: Dealing with massive resources in parallel infrastructures. In: Nicosia, G., Pardalos, P. (eds.) LION 7. LNCS, vol. 7997, pp. 444–449. Springer, Heidelberg (2013)
7. Laredo, J.L.J., Eiben, A.E., van Steen, M., Merelo, J.J.: On the run-time dynamics of a peer-to-peer evolutionary algorithm. In: Rudolph, G., Jansen, T., Lucas, S., Poloni, C., Beume, N. (eds.) PPSN X. LNCS, vol. 5199, pp. 236–245. Springer, Heidelberg (2008)
8. Miller, B.L., Goldberg, D.E.: Genetic algorithms, selection schemes, and the varying effects of noise. Evol. Comput. 4(2), 113–131 (1996)
9. Poli, R.: Tournament selection, iterated coupon-collection problem, and backward-chaining evolutionary algorithms. In: Wright, A.H., Vose, M.D., De Jong, K.A., Schmitt, L.M. (eds.) FOGA 2005. LNCS, vol. 3469, pp. 132–155. Springer, Heidelberg (2005)
10. Rost, B., Sander, C.: Combining evolutionary information and neural networks to predict protein secondary structure. Proteins 19(1), 55–72 (1994)
11. David Schaffer, J., Eshelman, L.J.: On crossover as an evolutionarily viable strategy. In: Belew, R.K., Booker, L.B. (eds.) ICGA, pp. 61–68. Morgan Kaufmann (1991)
12. Sokolov, A., Whitley, D.: Unbiased tournament selection. In: Proceedings of the 2005 Conference on Genetic and Evolutionary Computation, GECCO 2005, pp. 1131–1138. ACM, New York (2005)
13. Xie, H., Zhang, M.: Impacts of sampling strategies in tournament selection for genetic programming. Soft Comput. 16(4), 615–633 (2012)
14. Xie, H., Zhang, M., Andreae, P.: Another investigation on tournament selection: modelling and visualisation. In: Proceedings of the 9th Annual Conference on Genetic and Evolutionary Computation, GECCO 2007, pp. 1468–1475. ACM, New York (2007)

Diversity-Driven Selection of Multiple Crossover Operators for the Capacitated Arc Routing Problem

Pietro Consoli and Xin Yao

CERCIA,
School of Computer Science, University of Birmingham,
Birmingham, West Midlands, B15 2TT, UK
{p.a.consoli,x.yao}@cs.bham.ac.uk

Abstract. The Capacitated Arc Routing Problem (CARP) is a NP-Hard routing problem with strong connections with real world problems. In this work we aim to enhance the performance of MAENS, a state-of-the-art algorithm, through a self-adaptive scheme to choose the most suitable operator and a diversity-driven ranking operator. Experimental results on 181 problem instances show how these techniques can both improve the results of the current state-of-the-art algorithms and provide good directions to develop EAs with a more robust approximation ratio.

Keywords: Memetic Algorithm, Stochastic Ranking, Capacitated Arc Routing Problem, Self-Adaptation, Approximation Algorithms.

1 Introduction

The *Capacitated Arc Routing Problem* (CARP) 1981[12] has been an object of study mostly because of its close relationship with real world problems. For the CARP a considerable number of exact methods, heuristics and meta-heuristics has been created. Among the meta-heuristics, several population-based approaches can be mentioned. Lacomme et al. proposed a Genetic Algorithm[17], a Memetic Algorithm[18] as well as an ant-based approach[19]. Chu et al. proposed a scatter search in 2006[4] while Beullens et al. proposed a Guided Local Search [3] using a heuristic based on the evaluation of the potential mutations of each solution. Hertz and Mittaz[14] used a Variable Neighbourhood Search which uses a set of new mutation operators.

Tang et al. proposed a Memetic Algorithm[32] named MAENS whose local search is provided with a long-step operator, named Merge-Split.

Several algorithms have been proposed for different versions of the problem, such as the Memetic Algorithm in [25] for the Periodic CARP. Mei et al. also proposed an approach for the multi objective CARP in [24]. Another attempt based on a Tabu Search is described in [23]. Xing et al. developed in [36] a hybrid algorithm based on the ACO meta-heuristic for the Extended CARP, which introduces several constraints, such as a maximum service time, penalties

C. Blum and G. Ochoa (Eds.): EvoCOP 2014, LNCS 8600, pp. 97–108, 2014.
© Springer-Verlag Berlin Heidelberg 2014

for turns and a variable amount and position of depots. In [37] Xing et al. proposed an evolutionary approach for the Multi-Depot CARP. More recently, Mei et al[22] adopted a Cooperative Co-evolution framework[28] to decompose Large Scale CARP instances.

For NP-Hard problems strongly connected with real world applications as the CARP, it is important that algorithms that we design are reliable enough to return good solutions, when not optimal, in as many cases as possible. However, for this problem, the work on the reinforcement of the average results of the algorithms has been lacking.

The aim of this research is therefore to strengthen the approximation ability of a state-of-the-art algorithm for the CARP through the following techniques:

- the use of a novel distance measure between CARP solutions;
- a diversity-driven stochastic ranking operator;
- a set of new recombination operators for the problem;
- an Adaptive Operator Selection strategy which uses a new Credit Assignment technique based on the aforementioned ranking operator.

The rest of the paper is organized as follows. Section 2 introduces some preliminary notions such as the definition of the CARP and the MAENS algorithm which represents the state-of-the-art for this problem. Section 3 presents a novel diversity measure for CARP solutions and a diversity-driven ranking operator. Section 4 introduces a suite of crossover operators for the problem and the Adaptive Operator Selection technique adopted. Section 5 shows the performance of the MAENS algorithm when using the proposed techniques. Section 6 includes the conclusions and the future work.

2   Background

2.1   Problem Definition

$G = (V, A)$ is a connected directed graph where V is the set of vertices, A is the set of arcs, and the subset $A_R \subset A$ is the subset of *required arcs*. Elements of A are also called *tasks*, while A_R will be the *required tasks*. Each tasks t has a *demand* $d(t)$ which indicates the load necessary to serve the task, a *service cost* $sc(t)$ of crossing the task, a *dead-heading cost* $dc(t)$ of crossing the task without serving it, an *ID* and a reference to their *head(t)* and their *tail(t)* task. The *tasks* need to be served by a fleet of m vehicles with a capacity C and whose route starts and ends in a vertex called *depot*. Each task must be served within a single tour and each vehicle is bound to its capacity. A solution for a CARP instance can therefore be represented by a set of routes, which are sequences of tasks that need to be visited in the given order. To distinguish between routes a *dummy task* is commonly used with *ID* 0, which represents the vehicle being in the depot with null *service cost* and *demand*. The objective is therefore to minimize the total service cost (TC) of the routing subject to the previously mentioned constraints.

Table 1. Summary of CARP definition. $app(S_i)$ counts the times that a task appears in the sequence S, $inv(S_i)$ returns the task in the opposite direction of S_i and $nveh$ is the number of maximum vehicles allowed.

Problem Definition	$\min TC(S) = \sum_{i=1}^{\text{length}(S)-1}(sc(S_i) + sp(S_i, S_{i+1}))$
Constraints	$load(R_i) \leq Q$ $app(S_i) = 1, \forall S_i \in A_R$ $m <= nveh$
Total Service Cost	$TC = \sum_{i=1}^{\text{length}(S)-1}(sc(S_i) + sp(S_i, S_{i+1}))$
Total Load	$load(R_k) = \sum_{i=1}^{\text{length}(R_k)} d(S_{ik})$

The TC is calculated in the following way. When the vehicle is serving a *required task* S_i, the TC will include its serving cost $sc(S_i)$ plus the total cost of the shortest path (sp) necessary to connect the *tail* of the task to the *head* of the next task S_{i+1}, obtained through the use of Dijkstra algorithm[7]. A full definition of the problem is included in Table 1.

2.2 MAENS

The Memetic Algorithm with Extended Neighbourhood Search (MAENS)[32] is one of the most competitive and efficient algorithms in the context of the CARP. Its pseudo-code is provided in figure 1. It is possible to divide it into four phases: initialization (lines 1-7), crossover (lines 10-11), local search (line 14) and stochastic ranking (line 26). During the initialization phase, solutions are generated through the use of the Path-Scanning procedure [12]. In the main cycle, couples of parent solutions are randomly selected to generate an offspring using the Sequence Based Crossover (SBX) operator [29]. A local search is therefore performed on the neighbourhood of the solution, with a probability *lsprob*, using three move operators, namely Single Insertion (one task is moved to another route), Double Insertion (two consecutive tasks are moved to another route) and Swap (two tasks of different routes exchange their places). The best solution is then improved through the Merge-Split operator, which applies the Path

```
1.  initialise a population pop;
2.  while (pop not full OR attempts < trial)
    do
3.      generate a new individual p;
4.      if (p is not a clone) then
5.          add p to pop;
6.      end if
7.  end while
8.  while (stopping criterion is not satisfied)
    do
9.      for (i = 0 to size*offset) do
10.         randomly select p₁, and p₂ from
            pop;
11.         generate sᵢₓ = SBX(p₁, p₂);
12.         extract a random value r
13.         if (r < lsprob) then
14.             generate sᵢₗₛ = LocSearch(sᵢₓ,p);
15.         else
16.             sᵢₗₛ = sᵢₓ;
17.         end if
18.         if (sᵢₗₛ is a clone of a parent) then
19.             overwrite parent;
20.         else
21.             if (sᵢₗₛ not a clone) then
22.                 add to pop;
23.             end if
24.         end if
25.     end for
26.     pop = Stochastic Ranking(pop);
27. end while
```

Fig. 1. MAENS Algorithm

Scanning procedure[12] and the Ulusoy Splitting tour [34] to the tasks of two randomly selected routes. The local search procedure is completed by another iterative search using the three move operators. The offspring are then compared to the solutions of the previous generation and sorted using the Stochastic Ranking procedure [31].

Analysis of the Algorithm. We have analysed the algorithmic features of MAENS[32] in order to identify what might be the possible drawbacks that affect the robustness of its results. As a memetic algorithm, it is equipped with a local search operator which greatly improves its capacity of exploiting the good solutions found during the search. This exploitation ability might not be balanced enough by an efficient exploration as the algorithm does not have any mechanism that maintains the diversity of the population. The algorithm makes use of a single heuristic, the Path-Scanning[12], in both the initialization phase and within the local search. This might affect the ability of the algorithm to generate new routes during this phase. We choose therefore to contrast the exploitation ability of the algorithm embedding a control over the diversity of the population in the ranking operator and increasing the breadth of the recombination move replacing the SBX with a suite of different operators.

2.3   Approximation Algorithms

An Approximation Algorithm can be defined as follows[26]. Given a minimization optimization problem P with a cost function c, and an algorithm A that is able to produce a feasible solution $f_A(I)$ for the instance I of P whose optimal solution is $\overline{f}(I)$, then the algorithm A is an $\epsilon-$approximate algorithm of P if for any instance I of P :

$$\epsilon \geq \frac{|c(f_A(I)) - c(\overline{f}(I))|}{c(\overline{f}(I))}$$

for some value $\epsilon \geq 0$ and ϵ is the *approximation ratio* of A. As explained in [13], when evaluating the approximation of EAs, due to their stochastic nature, it would be more convenient to evaluate the estimated value $E(f_A(I))$ instead of the results obtained by a single execution of the EA.

3   A Distance Measure for the CARP

In order to be able to evaluate the performance of the algorithm in terms of exploration ability, it is necessary to define a distance measure between solutions. We have identified three possible approaches to define such a measure and we propose a novel one.

3.1   Measuring the Average Diversity of the Population

A first approach is to consider routes as clusters of tasks and consequently to choose an index that is commonly used in the data clustering context. Two non trivial issues affect this approach: it does not take into account the task service order and it has a high computational cost ($O(n^2)$).

One could choose a different approach by considering routes as strings and to use similarity indexes taken from this field, such as the Levenshtein distance [20]. Although this approach is more precise than the previous one, it is still computationally non trivial ($O(n^2)$) and it has the issue of depending on the chosen representation.

A third approach is that of considering the relationship of consecutiveness between edges as items of the dataset (e.g. if task t_2 follows task t_1 the couple (t_1, t_2) is an item of the dataset) and using a similarity measure between sets. The task service order is therefore split into couples of tasks. This approach has a linear computational time ($O(n)$) and it achieves a higher precision, if compared to the classic clustering approach, by taking into account the task service order. Such a measure has been successfully used in the context of multi-objective Vehicle Routing Problem as in [10] where the number of overlapping arcs is measured through the use of the Jaccard Index[16]. However, since this measure is not a proper metric, despite being valid to perform comparisons between solutions, it might be not fit to compute average distances. For this reason we propose a new distance measure able to deal with this issue.

3.2   A Revised Distance Measure Based on Neighbour Tasks

```
1.  find p1(t) and n1(t)∀t in S1;
2.  find p2(t) and n2(t)∀t in S2;
3.  for (each task t) do
4.      if (task t is served in both solutions)
        then
5.          if ( p1(t) = p2(t) ) then
6.              add one;
7.          end if
8.          if ( n1(t) = n2(t) ) then
9.              add one;
10.         end if
11.     end if
12. end for
13. divide the obtained value by 2N.
```

Fig. 2. Diversity Measure for CARP

Similarly to the aforementioned measure, we exploit the relation of consecutiveness between tasks inside the routes. However, the similarity measure adopted in this work is not based on the number of shared arcs between two tasks, but on the number of tasks which share their previous or next arcs. This approach maintains the linear computational cost and takes into account the service order, but has the advantage of always having consistent measurements as the number of tasks is a constant value. The proposed metric between two solutions S_1 and S_2 is described in Figure 3.2.

We define $p_i(t)$ and $n_i(t)$ as the functions that return respectively the previous and the next tasks of task t in solution i. Clearly this similarity measure will be equal to 0 if the two solutions are completely different (when $p_1(t) \neq p_2(t)$ and $n_1(t) \neq n_2(t), \forall t$) and equal to 1 if the solutions are identical ($p_1(t) = p_2(t)$ and $n_1(t) = n_2(t), \forall t$). We also point out as the $2N$ possible values achievable are uniformly distributed in the $[0-1]$ space, this allows the calculation of average similarities within the population.

3.3 Diversity-Driven Stochastic Ranking

We have initially used such a measure to define a new ranking operator with the aim to preserve the diversity of the population.

In order to do that, we have embedded this information in the stochastic ranking operator[31], which has also been used in the MAENS algorithm[32]. The pseudo-code of the Diversity-Driven Stochastic Ranking is included in Figure 3.3, where functions $rts(\cdot)$, $f(\cdot)$, $v(\cdot)$ refer respectively to the number of vehicles, the fitness function and the amount of violation of each solution and *maxrts* is the maximum number of available vehicles. A first condition limits the number of solutions violating the con-

```
1.  for (size of the population) do
2.      for (each consecutive couple p and q) do
3.          extract a random value r;
4.          switch p and q IF:
5.          a) (rts(p) < rts(q) and rts(p) > maxrts);
6.          b) (d_avg(p) < d_avg(q) and r < R_1);
7.          c) (f(p) > f(q) and r < R_2);
8.          d) (f(p) = f(q) and d_avg(p) < d_avg(q));
9.          e) (v(p) < v(q)).
10.     end for
11. end for
```

Fig. 3. Diversity-Driven Stochastic Ranking

straint relative to the maximum number of allowed vehicles. Thus, besides comparing the solutions according to their fitness and violation, we use the value d_{AVG} as a *crowding distance* to bias the search towards the areas of the search space which have not been thoroughly exploited. Therefore, we switch the position of the solutions also in the case of reduced diversity. As this is the first attempt to use such a measure, the work might be extended by using a more refined one that might consider either only the n closest solutions, or the number of solutions which are at least λ-distant or also the distance from the *centroid* individual of the population.

4 Operator Selection

As previously mentioned, the use of a single heuristic to generate routes might limit the exploration capacity of the algorithm and consequently its ability to escape local optima.

We propose the use of a suite of crossover operators under the assumption that it might lead to a more robust performance of the algorithm.

This offers the advantage to use each heuristic in those instances where they perform better. As this is not known a priori, it is necessary to define a operator-selection strategy which is able to address the problem of selecting the best performing crossover operator for each instance. Besides, the use of the proper combination of heuristics might improve further the performance of the single use of each one of them, as different heuristics might be the best in different phases of the search.

4.1 Crossover Operators

We have defined four new crossover operators for the CARP problem, namely GSBX, GRX, PBX, SPBX, which are described in this section.

Greedy Sequence Based Crossover (GSBX). The first operator applies a greedy selection to the SBX[29] operator. A route from each solution is randomly selected. These routes are therefore split in two parts and combined in order to generate two different solutions.

The greedy choice replaces the random selection of routes using the following rule that supports the selection of those routes which might have been less efficiently filled. Thus, the route A will have a higher probability to be chosen than route B if $Q - \text{load}(A) \leq Q - \text{load}(B)$ where Q is the maximum load that a vehicle can carry.

Greedy Route Crossover (GRX). The GRX operator adapts the concept of the GPX operator for the Graph Colouring Problem [9] in the context of the CARP. With a round robin criterion, the best route in one of the parent solutions is selected. The selection is performed with the same greedy-rule used in the GSBX operator.

The route is then copied in the offspring and its tasks are removed from parent solution routes. This process is repeated until all remaining routes R in parent solutions have $\text{load}(R) < Q/2$. The remaining routes are therefore randomly selected and combined to form new routes or inserted (when $|R| = 1$) in the existing ones in the positions which minimize the total service cost of the solution.

Pivot Based Crossover (PBX). The PBX operator ranks a list of tasks in a similar way to the Augment-Insert[27] heuristic, although it differs from it as the tasks are ranked according to the load of their outward and return paths, instead of their total service cost.

A route from each parent individual is randomly selected and their tasks are inserted in a unordered list L. A *pivot* task is therefore identified by choosing with higher probability if either its outward or return path has the highest load when using only tasks $t \in L$. The selected subpath is adopted and the rest of the route is built by choosing to insert the tasks in L that minimize the total cost of the route and that do not violate the capacity constraint.

Shortest Path Based Crossover (SPBX). The SPBX applies the idea of the PBX operator to a couple of tasks (t_A, t_B). If the total cost of the shortest path using the tasks of a list L is $SP_L(\cdot, \cdot)$, the algorithm will select with a higher probability the couple t_A and t_B belonging to L such that $SP_L(t_A, t_B)$ is maximised while $SP_L(depot, t_A)$ and $SP_L(t_B, depot)$ are minimised.

4.2 Adaptive Operator Selection

The Adaptive Operator Selection (AOS) is typically composed of an *Operator Selection Rule* (OSR) , such as [11,33], and a *Credit Assignment Mechanism*.

The Operator Selection scenario can be seen as a dynamic version of the game theory Multi Armed Bandit problem. Each operator represents an arm with an unknown reward probability. The aim is therefore to select during each time-step the arm that maximises the probability of obtaining the reward. Several experiments have shown how MAB-based approaches outperform the former techniques[5].

Common Credit Assignment techniques rely on the evaluation of the fitness function, assigning a higher reward to the operator whose offspring shows the higher improvement. Some reward techniques which also consider diversity have been proposed in the context of multi-modal problems, such as [21]. We have chosen therefore to adopt a MAB-based technique, called dMAB, first proposed in [5], and we have combined it with a new credit assignment technique based on the use of the diversity-driven ranking operator that we have previously introduced. The choice of dMAB is merely dictated by the fact that it is one of the most promising techniques in the area of AOS, as in this stage we are not interested into identifying the most efficient operator. Comparisons of the performance of dMAB with respect other techniques, which have not been performed in this context, can be found in dMAB original paper[5].

Dynamic Multi-Armed Bandit. The dMAB adapts the classic Multi-Armed Bandit scenario to a dynamic context where the reward probability of each arm is not independent and is not fixed. To address the dynamic context problem, the classical Upper Confidence Bound (UCB1)[1] algorithm is combined with a Page Hinkley test[15], to identify the change of reward probabilities. More information can be found in the original paper [5].

Credit Assignment by Stochastic Ranking. We propose a new credit assignment mechanism exploiting the selection operated by the diversity-driven ranking operator. We assign a reward r which is proportional to the number of offspring generated by the selected operator that will survive to the next generation after the stochastic ranking operator has been applied. Therefore, $r = 0$ when none of the individuals generated by the selected operator has survived the ranking process and $r = 1$ when only individuals generated by it have been chosen by the ranking operator.

Such a technique shows several advantages with respect to classic fitness-based credit assignment ones:

- it takes into account the fitness of the solutions, their similarity and the violation of the constraints;
- the adaptive operation selection process does not require domain knowledge;
- the reward values are always normalized and there is no need to derive a scaling factor;
- unlike fitness-based techniques, it is not affected by the convergence speed of the algorithm.

5   Experimental Studies

We have tested the results of the original algorithm against several versions that we have labelled *MAENSd*, *MAENSm* and *MAENS**, which are respectively the versions of the algorithm adopting the diversity-driven stochastic ranking

Table 2. MAENS* parameters

Name	Description	Value
psize	population size	30
ubtrial	maximum attempts to generate each initial solution	50
opsize	offspring generated during each generation	6*psize
P_{ls}	probability of performing the local search	0.2
p	routes selected during MergeSplit	2
G_m	maximum generations	500
SR_{r1}	probability of sorting solutions according to their diversity	0.25
SR_{r2}	probability of sorting solutions according to their fitness	0.7
σ	tolerance factor for Page-Hinkley test	0.05
λ	change threshold for Page-Hinkley test	1.25

operator, the one using the proposed AOS strategy and the combination of both techniques.

The comparison has been carried out on four benchmark test sets, namely gdb[6] (23 instances), val[2] (34 instances), egl[8] (24 instances), and *Beullens et al.*[3], which is composed of four groups of 25 instances, namely C,D,E and F.

The Wilcoxon Signed-Rank test[35] has been used to perform a statistical hypotesis test between MAENS and each of the proposed versions. The test has been conducted using the R software environment[30]. In each case, the test has rejected the null hypothesis that the results of the two algorithms were not significantly different. Table 4 reports the details of such tests.

Table 3. A summary of the results of the four algorithms. Each column shows the number of instances where each algorithm achieved a better average fitness (W), performed equally (D) or worse (L) than MAENS, the mean average fitness (avg), standard deviation (std) and best result (best) as well as the mean approximation ratio (ϵ).

	MAENS	MAENSd	MAENSm	MAENS*
W	–	85	79	91
D	–	76	74	77
L	–	20	28	13
avg	2040.77	2036.05	2035.90	2034.15
std	9.42	5.23	6.37	5.02
best	2026.25	2028.06	2026.09	2026.24
ϵ	.0195	.0161	.0164	.0154

Table 2 shows the parameters used to execute the algorithm for 30 independent trials on each of the 181 instances. The algorithm has not been through a process of parameter configuration. Parameters present in the original version of the algorithm (first 7 parameters included in table 2) have kept their original values. New parameters such as those necessary for the Page-Hinckley test have been identified through a few test-and-trial attempts, using the set of benchmark instances as a training set.

For the sake of brevity we do not include the complete results of the comparisons, which are however available on the authors website[1], but we only report a summary of the results in table 3. In terms of mean average fitness, both MAENSd and MAENSm manage to outperform the original algorithm in 85 and 79 instances, and lose the comparison only in 20 and 28 cases. The

Table 4. Results of two-sided tests of significance for the three proposed versions of the MAENS algorithm. The columns show the V statistic computed with the Wilcoxon Signed-Rank Test and the p-value obtained.

	MAENSd	MAENSm	MAENS*
V	4773	4917	5039
p-values	1.995e-10	2.945e-10	7.096e-14

[1] http://www.cs.bham.ac.uk/~pac265/

results of their combined version MAENS* confirm how the combination of the two techniques has a constructive effect on the algorithm, achieving a better average fitness in 91 instances and losing in only 13. In terms of average standard deviation, the results of MAENSd and MAENSm represent an improvement with respect to the original algorithm, lowering it from 9.42 to 5.23 in the first case and to 6.37 in the second case. Even in this case the results of MAENS* improve the result achieving an average standard deviation of 5.02. We interpret this result as a sign of an improved convergence reliability. The average mean results obtained by the three algorithms reflects the previously mentioned analysis.

An interesting result is that of the average best result obtained by the algorithms. MAENSd showed a somehow predictable reduction of the algorithm exploitation ability (2028.06 against 2026.25 of MAENS) while MAENSm managed to slightly improve the average best result (2026.09). MAENS* coherently achieved results between those of MAENSd and MAENSm (2025.82), confirming the effectiveness of this combination, which discovered new optima for 18 instances.

We have also compared the algorithms in terms of their average approximation-ratio. MAENS ratio of 0.0195 has been reduced to 0.0160 and 0.0164 in the cases of MAENSd and MAENSm and to 0.0154 in the case of MAENS*.

With regard to the runtime, MAENS* is essentially comparable to its original version. The additional computational cost introduced by the calculation of the average diversity of the solutions is balanced by the improved convergence speed, while the AOS does not add any noticeable cost in the algorithm, whose computational cost is still largely dominated by the local search procedure.

6   Conclusions

We have proposed an improved version of the current state-of-the-art algorithm for the CARP[32], called MAENS*. The main characteristics of this algorithm are (a) a new diversity measure between solutions, (b) a diversity-driven stochastic ranking operator, (c) an AOS strategy using four novel crossover operators and (d) a novel Credit Assignment strategy using the aforementioned ranking operator to define rewards. The results of a comparison between the two algorithms show how MAENS* outperformed the the original algorithm in terms of average fitness and produced more robust and reliable results.

The work carried out so far leaves space to several possible improvements, as optimal values of the parameters adopted could be identified, as well as generalizations of the techniques proposed in this work to other combinatorial optimisation problems.

Acknowledgement. This work was supported by EPSRC (Grant No. EP/ I010297/ 1). Xin Yao was supported by a Royal Society Wolfson Research Merit Award. The authors would like to thank the anonymous reviewers for their insightful and constructive comments.

References

1. Auer, P., Cesa-Bianchi, N., Fischer, P.: Finite-time analysis of the multiarmed bandit problem. Machine Learning 47(2-3), 235–256 (2002)
2. Benavent, E., Campos, V., Corberán, A., Mota, E.: The capacitated arc routing problem: lower bounds. Networks 22(7), 669–690 (1992)
3. Beullens, P., Muyldermans, L., Cattrysse, D., Van Oudheusden, D.: A guided local search heuristic for the capacitated arc routing problem. European Journal of Operational Research 147(3), 629–643 (2003)
4. Chu, F., Labadi, N., Prins, C.: A scatter search for the periodic capacitated arc routing problem. European Journal of Operational Research 169(2), 586–605 (2006)
5. Da Costa, L., Fialho, A., Schoenauer, M., Sebag, M., et al.: Adaptive operator selection with dynamic multi-armed bandits. In: Genetic and Evolutionary Computation Conference (GECCO), pp. 913–920 (2008)
6. DeArmon, J.S.: A comparison of heuristics for the capacitated Chinese postman problem. Ph.D. thesis, University of Maryland (1981)
7. Dijkstra, E.W.: A note on two problems in connexion with graphs. Numerische Mathematik 1(1), 269–271 (1959)
8. Eglese, R.W.: Routing winter gritting vehicles. Discrete Applied Mathematics 48(3), 231–244 (1994)
9. Galinier, P., Hao, J.K.: Hybrid evolutionary algorithms for graph coloring. Journal of Combinatorial Optimization 3(4), 379–397 (1999)
10. Garcia-Najera, A.: Preserving population diversity for the multi-objective vehicle routing problem with time windows. In: Proceedings of the 11th Annual Conference Companion on Genetic and Evolutionary Computation Conference: Late Breaking Papers, pp. 2689–2692. ACM (2009)
11. Goldberg, D.E.: Probability matching, the magnitude of reinforcement, and classifier system bidding. Machine Learning 5(4), 407–425 (1990)
12. Golden, B.L., Wong, R.T.: Capacitated arc routing problems. Networks 11(3), 305–315 (1981)
13. He, J., Yao, X.: An analysis of evolutionary algorithms for finding approximation solutions to hard optimisation problems. In: The 2003 Congress on Evolutionary Computation, CEC 2003, vol. 3, pp. 2004–2010. IEEE (2003)
14. Hertz, A., Mittaz, M.: A variable neighborhood descent algorithm for the undirected capacitated arc routing problem. Transportation Science 35(4), 425–434 (2001)
15. Hinkley, D.V.: Inference about the change-point from cumulative sum tests. Biometrika 58(3), 509–523 (1971)
16. Jaccard, P.: Etude comparative de la distribution florale dans une portion des Alpes et du Jura. Impr. Corbaz (1901)
17. Lacomme, P., Prins, C., Ramdane-Chérif, W.: A genetic algorithm for the capacitated arc routing problem and its extensions. In: Boers, E.J.W., Gottlieb, J., Lanzi, P.L., Smith, R.E., Cagnoni, S., Hart, E., Raidl, G.R., Tijink, H. (eds.) EvoWorkshop 2001. LNCS, vol. 2037, pp. 473–483. Springer, Heidelberg (2001)
18. Lacomme, P., Prins, C., Ramdane-Cherif, W.: Competitive memetic algorithms for arc routing problems. Annals of Operations Research 131(1-4), 159–185 (2004)
19. Lacomme, P., Prins, C., Tanguy, A.: First competitive ant colony scheme for the CARP. In: Dorigo, M., Birattari, M., Blum, C., Gambardella, L.M., Mondada, F., Stützle, T. (eds.) ANTS 2004. LNCS, vol. 3172, pp. 426–427. Springer, Heidelberg (2004)

20. Levenshtein, V.I.: Binary codes capable of correcting deletions, insertions and reversals. Soviet Physics Doklady 10, 707 (1966)
21. Maturana, J., Saubion, F.: A compass to guide genetic algorithms. In: Rudolph, G., Jansen, T., Lucas, S., Poloni, C., Beume, N. (eds.) PPSN X. LNCS, vol. 5199, pp. 256–265. Springer, Heidelberg (2008)
22. Mei, Y., Li, X., Yao, X.: Cooperative co-evolution with route distance grouping for large-scale capacitated arc routing problems. IEEE Transactions on Evolutionary Computation (accepted on July 31, 2013)
23. Mei, Y., Tang, K., Yao, X.: A global repair operator for capacitated arc routing problem. IEEE Transactions on Systems, Man, and Cybernetics, Part B: Cybernetics 39(3), 723–734 (2009)
24. Mei, Y., Tang, K., Yao, X.: Decomposition-based memetic algorithm for multi-objective capacitated arc routing problem. IEEE Transactions on Evolutionary Computation 15(2), 151–165 (2011)
25. Mei, Y., Tang, K., Yao, X.: A memetic algorithm for periodic capacitated arc routing problem. IEEE Transactions on Systems, Man, and Cybernetics, Part B: Cybernetics 41(6), 1654–1667 (2011)
26. Papadimitriou, C.H., Steiglitz, K.: Combinatorial optimization: algorithms and complexity. Courier Dover Publications (1998)
27. Pearn, W.L.: Augment-insert algorithms for the capacitated arc routing problem. Computers & Operations Research 18(2), 189–198 (1991)
28. Potter, M.A., De Jong, K.A.: A cooperative coevolutionary approach to function optimization. In: Davidor, Y., Männer, R., Schwefel, H.-P. (eds.) PPSN III. LNCS, vol. 866, pp. 249–257. Springer, Heidelberg (1994)
29. Potvin, J.Y., Bengio, S.: The vehicle routing problem with time windows part ii: genetic search. INFORMS Journal on Computing 8(2), 165–172 (1996)
30. R Core Team: R: A Language and Environment for Statistical Computing. R Foundation for Statistical Computing, Vienna, Austria (2013), http://www.R-project.org
31. Runarsson, T.P., Yao, X.: Stochastic ranking for constrained evolutionary optimization. IEEE Transactions on Evolutionary Computation 4(3), 284–294 (2000)
32. Tang, K., Mei, Y., Yao, X.: Memetic algorithm with extended neighborhood search for capacitated arc routing problems. IEEE Transactions on Evolutionary Computation 13(5), 1151–1166 (2009)
33. Thierens, D.: An adaptive pursuit strategy for allocating operator probabilities. In: Proceedings of the 2005 Conference on Genetic and Evolutionary Computation, pp. 1539–1546. ACM (2005)
34. Ulusoy, G.: The fleet size and mix problem for capacitated arc routing. European Journal of Operational Research 22(3), 329–337 (1985)
35. Wilcoxon, F.: Individual comparisons by ranking methods. Biometrics Bulletin 1(6), 80–83 (1945)
36. Xing, L.N., Rohlfshagen, P., Chen, Y.W., Yao, X.: A hybrid ant colony optimization algorithm for the extended capacitated arc routing problem. IEEE Transactions on Systems, Man, and Cybernetics, Part B: Cybernetics 41(4), 1110–1123 (2011)
37. Xing, L., Rohlfshagen, P., Chen, Y., Yao, X.: An evolutionary approach to the multidepot capacitated arc routing problem. IEEE Transactions on Evolutionary Computation 14(3), 356–374 (2010)

Dynamic Period Routing
for a Complex Real-World System:
A Case Study in Storm Drain Maintenance

Yujie Chen, Peter Cowling, and Stephen Remde

York Centre for Complex Systems Analysis (YCCSA)
and Dept. of Computer Science, University of York, UK
{yc1005,peter.cowling,stephen.remde}@york.ac.uk

Abstract. This paper presents a case study of a real world storm drain maintenance problem where we must construct daily routes for a maintenance vehicle while considering the dynamic condition and social value of drains. To represent our problem, a dynamic period vehicle routing problem with profit (DPVRPP) model is proposed. This differs from the classical period routing problem in a number of ways. Firstly, it is dynamic: during the planning horizon, the demands from damaged drains and residents reports arrive continuously. In addition, the drains condition is changing over time. Secondly, our objective is maximizing the profit, defined here as the drains condition with respect to its social value.

This study is based on large-scale data provided by Gaist Solutions Ltd. and the council of a UK town (Blackpool). We propose an adaptive planning heuristic (APH) that produces daily routes based on our model and an estimation of changing drain condition in the future. Computational results show that the APH approach can, within reasonable CPU time, produce much higher quality solutions than the scheduling strategy currently implemented by Blackpool council.

1 Introduction

This paper describes a study of scheduling vehicle routes for the drain maintenance problem of Blackpool. The storm drain maintenance system in Blackpool records 28,290 drains distributed over approximately 36.1km^2. Drain cleaning is undertaken by Blackpool local council who presently deploys one team with 2 vehicles, one for daily cleaning and a high pressure jet machine for blocked drain cleaning. Due to the limitation of the number of workers, only one vehicle works each day. Every day the vehicle should visit and clean the planned drains and return to the depot. Usually, drains are cleaned once per year.

In addition to the normal daily cleaning, the drain cleaning crews are also responsible for emergency events (e.g. residents calling reports or damage repair). Every drain has an associated social value. This value is derived from a number of social factors provided by Blackpool council. To be more specific, all the drains are graded between 0-6 by considering: which road they belong to; is it near a school; is it near a hospital; is it near an economic center or is it in

C. Blum and G. Ochoa (Eds.): EvoCOP 2014, LNCS 8600, pp. 109–120, 2014.

the district center. A higher social value implies that if this drain is blocked and floods happen here, it brings relatively larger economic and social losses. In other words, we prefer to clean the drains with higher social value more frequently to keep them working properly.

Based on the feature of the drain maintenance problem, we propose a new model called the dynamic period vehicle routing problem with profit (DPVRPP) and use an adaptive planning method to solve this problem. There are two main differences to the well-studied period vehicle routing problem (PVRP)[1]. First, there is no hard visit frequency constraint for each drain during the planning period. The condition of each drain is changing during the long planning period and emergency events may occur. Hence, the updated plan is activated in response to these changes. Second, instead of minimizing the total distance travelled, the DPVRPP aims to maximize the total profit during the planning horizon. In this case, the profit is measured by all the drains' condition with respect to their social value. Efficient routes that help to shorten the total travel distance are implicitly preferred.

The remainder of this paper is organised as follows. In Section 2, relevant literature is surveyed. Then, we present a mathematical model of the storm drain maintenance problem in Section 3, followed by the adaptive heuristic in Section 4. Section 5 reports a series of experiments and analysis. Section 6 gives the conclusions.

2   Related Works

The closest well-studied variation of the vehicle routing problem (VRP) is the PVRP [1]. As a generalization of the ordinary VRP problem, the PVRP extends the planning period from one day to many days. Over the planning period, each customer must be visited on a required number of days. In classical PVRP, customers' requirements are known in advance. They can be served in a set of certain patterns. For example, a customer requiring two visits per week can be arranged in Monday-Thursday or Wednesday-Saturday pattern. Therefore, the PVRP is also named as the allocation-routing problem [2].

A number of solutions for the PVRP are proposed since the problem appeared. Chao et al. [3] presented a two-phase record-to-record algorithm that is constructed by several local moves applied one after another for the PVRP. Cordeau et al. [4] and Brandao and Mercer [5] implemented a tabu search heuristic for the PVRP with time window constraints and heterogeneous vehicles. Alegre et al. [6] applied a scatter search framework to solve the PVRP with a very long planning period. This algorithm is designed based on a problem of assigning calendars to customers in a periodic vehicle loading problem [7]. A variable neighbourhood search (VNS) approach for the PVRP has been proposed by Hemmelmayr et al. [8]. Based on the work in [8], Pirkwieser and Raidl added a coarsening and refinement process into VNS, which is called multilevel VNS for the PVRP [9].

For our problem, one of the important characteristics is its profit maximization objective. Similar work has been done in the literature. Baptista et al. modelled a paper recycling problem as the PVRP with some new features [10], in which the visit frequency for each customer was not fixed. Instead, the number of visits within the planning period becomes a decision variable. Furthermore, their objective function adds the customer demand as profit. Gonalves et al. considered an oil collection problem [11]. They aimed to maximize the profit rather than minimize the travel distance during the period. Also, a visit frequency was treated as a decision variable which is related to the profit.

In terms of the dynamic routing problem, the literature is relatively sparse. Angelelli et al. simplified a dynamic multi-period vehicle routing problem (DM-PVRP) such that each request has to be served within two days and only one vehicle without capacity constraint is available [12]. As an extension of the previous work, a fleet of vehicles and on-line requests are considered using replanning during the day [13]. Wen et al. modelled a real-life dynamic multi-period and multi-objective routing problem and proposed a three-phase heuristic embedded within a rolling horizon scheme [14].

3   Storm Drains Maintenance Problem

The storm drain maintenance problem is considered as a long period dynamic routing problem. All the drains are scheduled in a daily maintenance schema, and calling reports may arrive at any time and have to be scheduled as early as possible but no hard constraints are given. In addition, not every drain is accessible when visited, usually caused by vehicles parked on top of them. Another situation is that a drain may be reported as damaged or not working properly, which requires a specific type of vehicle to fix. A set of heterogeneous vehicles provide the services. Every day, one vehicle departs from the depot at 9:00 am and returns at 5:00 pm. There are no other capacity constraints for the vehicles. This problem is dynamic because the drains' condition changes over time, calling reports are continuously arriving and damaged drains are increasing over time. Hence, we must re-plan the schedule with up to date information.

The drain maintenance problem is modelled as an undirected complete graph $G = (V, E)$, where V is the set containing N number of drains and one depot, and E is the set of edges $(i, j), \forall i, j \in V$. $c_{i,j}$ represents the travel time between vertex i and vertex j, where $\forall i, j \in V$. For each day, the planning process can be solved as the PVRP of the next r days. We describe the formulation for the scheduling problem on day t ($t \in D$, where D is the set of days in the whole scheduling period and $r \ll |D|$). Two types of vehicles M_{type} are provided: type one for daily maintenance and normal calling reports; type two for fixing damaged drains. Every day, the vehicle's maximum working time is T_{max}. t_1 and t_2 are time required to clean a normal drain and fix a damaged drain respectively. Furthermore, each drain has a social value Q_i, where larger number means this drain is more important or if it is blocked, it would cause worse consequences.

In order to maximise the overall maintenance quality under a dynamic environment, we need a method to estimate the condition of each drain over the planning horizon r. The Weibull distribution [15], a common predictor of equipment failures, with shape parameter m and scale parameter x_0 is used to estimate this.

$P_i(d)$ is the probability drain i is blocked on day d of the scheduling period

$$x_{idl} = \begin{cases} 1, \textit{if drain } i \textit{ is visited by type } l \textit{ vehicle on day } d \\ 0, \textit{otherwise} \end{cases}.$$

$$y_{ijdl} = \begin{cases} 1, \textit{if edge } (i,j) \textit{ is visited by type } l \textit{ vehicle on day } d \\ 0, \textit{otherwise} \end{cases}.$$

Using these notations, the drain maintenances problem on day t can be formulated as follows:

$$max \sum_{d=t}^{t+r-1} \sum_{i=1}^{N} Q_i(1 - P_i(d)) \tag{1}$$

Subject to:

$$\sum_{i=1}^{N} x_{idl}t_l + \sum_{(i,j) \in E} c_{i,j}y_{ijdl} \le T_{max}; \qquad \forall d \in D, l = 1,2 \tag{2}$$

$$x_{idl} = 1 \Leftrightarrow \exists! j, j' : y_{ijdl} = y_{j'idl} = 1; \qquad \forall d \in D, l = 1,2 \tag{3}$$

$$(\sum_{i=1}^{N} x_{id1})(\sum_{i=1}^{N} x_{id2}) = 0; \qquad \forall d \in D \tag{4}$$

$$x_{0dl} = 1; \qquad \forall d \in D, l = 1,2 \tag{5}$$

$$y_{0jdl} = 1; \qquad \forall d \in D, l = 1,2 \tag{6}$$

$$y_{i0dl} = 1; \qquad \forall d \in D, l = 1,2 \tag{7}$$

Objective function 1 maximizes the overall profit during the planning horizon from $d = t$ to $d = t + r - 1$. It is measured by each drain's condition $1 - P_i$ on each day multiplied by its social value. Fig. 1 a) illustrates the estimation of the probability of a drain i blocked since the last clean under damaged, calling report received and normal state, respectively. According to the drain cleaning life cycle data provided by Blackpool council, we set the shape parameter $m = 2$ in the Weibull function. The scale parameter x_0 equals 365, 20, 10 on *normal*, *calling* and *damaged* state respectively. Fig. 1 b) presents an example of the probability of one drain getting blocked changing over time. In the example, since the last cleaning day, the probability of getting blocked increases slowly over time. On the 100th day, this drain is recorded as damaged. Hence, from day 100 to day 120 the drain is blocked and then it is fixed on 120th day. On the 365th day, it is cleaned for the second time.

Constraint 2 guarantees that the time spent on each daily route is less than the limit T_{max}. Constraints 3, 5, 6 and 7 ensure every route is feasible and starts from depot and returns to the depot. Constraint 4 allows only one vehicle to work each day.

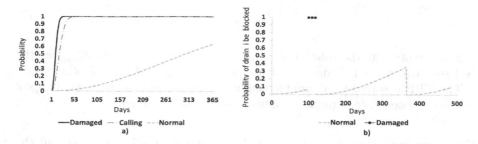

Fig. 1. a) Probability of drain i blocked in different states; b) An example of drain i' blocked probability changing over time

4   Adaptive Planning Heuristic (APH)

In order to react to emergency events and the changing situations, we propose a dynamic planning process at the beginning of each day. To avoid short-sighted plans that do not consider the overall drains condition over a longer period and to reduce the complexity of planning over a whole D day period, this method produces a future r days plan based on the most recent drain information.

Another issue we have to face is the large number of drains. Planning for the entire problem everyday is inefficient. Therefore, a two stage method is introduced, which includes routing and adaptive planing heuristics. For each stage, VNS is used as a search framework embedded with different local moves. This method is summarized in Algorithm 1.

Algorithm 1. Two stage routing and adaptive planing heuristic

---

Routing stage: Construct a set S_{basic} of optimized routes that visit all the drains at least once.
Scheduling stage:
for each day t in the time horizon D do
    Generate and update the set of new routes by considering up to date drains' information: S_{new}
    $S_{all} = S_{basic} \cup S_{new}$
    Generate the schedule for future r days by choosing r routes from S_{all}.
end for

In the adaptive planing heuristic, the parameter r is a user-defined parameter and depends on the specific problem. A bigger value of r considers a longer view of the entire scheduling horizon. In Section 5, an analysis on r is presented.

4.1   Routing Stage

The routing stage aims to find a group of optimized candidate routes that may be scheduled in the future days. In this stage, our objective function 1 is not applied. Instead, we purely treat it as a VRP using the objective function:

$$min \sum_{i \in V} \sum_{j \in V} c_{i,j} z_{i,j} \qquad (8)$$

S_{basic} represents the solution generated in the first stage and S_{basic}^* as the best solution. $z_{i,j} = 1$ if edge (i, j) is in solution S_{basic} and $z_{i,j} = 0$ otherwise.

The VRP is solved by a variable neighbourhood search heuristic made up of three components: construction, routes optimization, refining.

Construction. Due to the large number of drains, firstly we cluster all the drains by looking at which ward each drain belongs to. Blackpool is divided by 21 similar size wards. For each group of drains, the Clarke-Wright (CW) Savings heuristic [16] is used to solve the sub-problem as a VRP.

The possibility of inaccessible drains is considered during the process of route construction by using merge criteria 9.

$$n(1 - p_{inacc})t_1 + \sum_{(i,j) \in s} c_{i,j} \leq T_{max} \qquad (9)$$

Where n is the number of drains in the route $s \in S_{basic}$ and p_{inacc} is the probability of each drain being inaccessible when visited. Furthermore, the stop criteria of the CW algorithm used here is that no more routes can be merged without breaking the T_{max} limit. At the end of the CW algorithm we can not guarantee every route reaches the T_{max} or comes close to it.

Optimization. After an initial solution is constructed, an optimization phase is executed to improve it. So far, all the drains are included in solution S_{basic} with visiting frequency equal to 1. In the next step, we use the VNS with i-relocate and j-cross-exchange (see Fig. 2) to optimize all the routes in S_{basic} regardless of which ward the route belongs to. Here, i-relocate means that the segments changed in one relocated move have maximum number of nodes equals i. Similar, j-cross-exchange means that the segments changed in one cross-exchange move have maximum number of nodes equals j. Standard VNS [17] with first improvement acceptance is used. In total, 12 neighbourhoods are implemented. The order of neighbourhoods is i-relocate ($i = 1, 2, 3, 4, 5, 6$) and then j-cross-exchange ($j = 1, 2, 3, 4, 5, 6$).

In order to enhance the solution quality, a local search strategy is used after a solution is obtained through "shaking". Two of the most popular iterative improvement procedures 2-opt and 3-opt are applied here. 2-opt and 3-opt are single route improved strategies. Hence, only the two modified routes have to be re-optimized.

Refine Phase. At the end of the optimization phase, we should be able to get the optimum solution of the VRP. However, we can not guarantee every route maximizes the use of the daily time limitation T_{max}. Therefore, an insertion heuristic is implemented. For each route $s \in S_{basic}$, we try to insert more drains

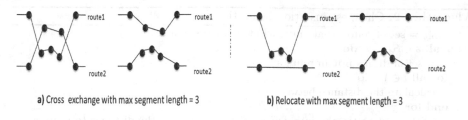

a) Cross exchange with max segment length = 3   **b) Relocate with max segment length = 3**

Fig. 2. a) Cross exchange; b) Relocated

which are not already in s using the Nearest Neighbour heuristic[18] until no more drains can be inserted without breaking the T_{max} limitation. Finally, 2-opt and 3-opt is applied to each $s \in S_{basic}$ to guarantee every s reaches its best value and we get our best solution for the preparation stage S_{basic}^*.

4.2   Adaptive Planning Stage

The adaptive planning process is run every day and makes the decision based on up to date information and the estimation of the future change. Two processes are implemented: updating the candidate routes set and scheduling.

Update Candidate Routes Set. At the end of each day, a new calling reports and b damaged drains reports are simulated. Calling reports that have not solved from previous days and a new calling reports from today make up the set V_{calls}. In the same way, we also get a set V_{damage}.

We use $S_{calling}$ to represent the set of routes that include drains in set V_{calls}. $S_{calling}$ can be obtained by using the method proposed in the routing stage without the refine step. Then, for each route $s \in S_{calling}$, we treat it as an opportunity to clean more drains near by. Hence, instead of the refine step presented in stage one, a variation of cheapest insertion procedure is shown here (see algorithm 2). To keep the algorithm simple, we measure the distance between drain i and route s by using the distance between i and the closest drain j on route s.

Because damaged drains should be handled by a different type of vehicle, we reconstruct the routes for all drains in the set V_{damage} by considering it as a VRP. The method used here is the CW heuristic [16]. Then we get a route collection S_{damage}.

At this point, we update our routes collection using the following rules:

$$S_{all} = S_{basic}^* + S_{calling} + S_{damage}$$

Scheduling Routes Using VNS. The final step of our process aims to produce an r-days schedule based on the optimized candidate routes set S_{all}. Firstly, we build an initial plan by choosing r number of routes with the highest $\sum_{i \in r} Q_i *$ $(1 - P_i(today))$. Then, a VNS method with i-replace and j-exchange (see Fig. 3) local moves is implemented to improve the initial plan. The VNS uses neighbourhood i-replace ($i = 2, 1$) then j-exchange ($j = 4, 3, 2, 1$) in order.

Algorithm 2. Cheapest insertion heuristic

$S_{calling}$ = set of routes that only include the drains in set V_{calls}
for all $s \in S_{calling}$ **do**
  V_s= set of drains **not** in route s
  for all $i \in V_s$ **do**
    calculate the distance between i and s
  end for
  sortlist= sort vertices in increasing order by means of the distance to route s
  repeat
    insert vertex i from the top of sortlist into the best position of route s
    delete i from sortlist
  until no more vertices in sortlist can be added to s without break T_{max} limit
end for
return $S_{calling}$

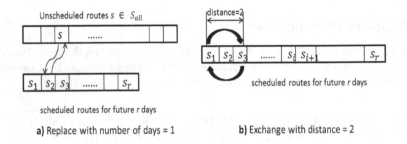

a) Replace with number of days = 1   b) Exchange with distance = 2

Fig. 3. a) Replace local move with one day schedule changed; b) Exchange local move with two days schedules changed

5   Computational Results

Data & Parameters. The drain location data in our research is provided by Blackpool council. On average 1500 calls are received every year. In our simulation, the probability of a drain to be recorded as a calling event relates to its condition $1 - P_i(d)$. The worse the drain condition, the higher probability a call is received. Here, we set the probability of receiving a call for drain i equal to $\gamma * P_i(d)$. γ is set into 0.0003 to match the real calling reports receiving situation. Since we do not have official data of the probability for the damage report and inaccessible drains, we approximate these values from Manchester council data. According to their records, 5.1% of drains do not get attended and 10.8% of drains get damaged per year. Due to the memory limitation, half of the drains chosen at random have been studied in the following experiments.

Planning Horizon Analysis. First, we investigate the effects of the future days planning parameter r to our solution quality. We test the APH with r from 1 to 9. The results are shown in Fig. 4 and we observe $r = 6$ produces the best plans. As we mentioned before, a bigger value of r considers a longer view

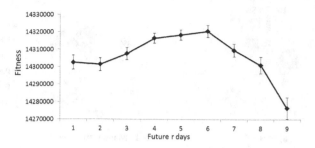

Fig. 4. The effect of parameter r

of the entire scheduling horizon. But, in the dynamic situation, too big r may mis-predict the future situation, which results in worse scheduling.

Comparison to Manual Strategy. Currently, a reactive strategy is used in Blackpool. Drains on the main street are scheduled into a annual cleaning schema. If any call or damaged report is received, the maintenance team will be taken away from the normal daily maintenance.

To compare the APH with the manual strategy (MS), thirty random runs using the manual strategy and the APH with $r = 6$ are performed. Table 1 and Fig. 5 summarize the result by comparing the average of drains' condition over one year. Over 76% drains have block probability less than 40% by using the APH. In comparison, this number of the MS is 28%. Fig. 6 illustrates the detail change over one year. Both the APH and the MS cases start with the same initial drains' situation. When we apply the APH approach, the overall drains condition gradually improves through the first half year and stays in a generally good condition. The reactive manual strategy shows a vicious circle: a bad strategy leads to worse drain condition; worse drain condition leads to more reports and increases the reactive maintenance.

Table 1. Performance of the APH and the MS with 95% confidence intervals

	APH	Manual
Normalized Fitness	158.24 ± 0.054	100 ± 0.903

In terms of social value, the APH also produces a significantly better solution compared to the manual strategy. Table 2 depicts the drains condition distribution with different social values. The APH always prefers to keep the higher social value drains in a good condition. Over 91% drains with social value = 6 have the block probability less than 40%.

Next we analyse the waiting time between a call being made and the drain being cleaned. Results show that when we use the APH approach, the total calls received through whole year is about 3 times less than using the MS. This is

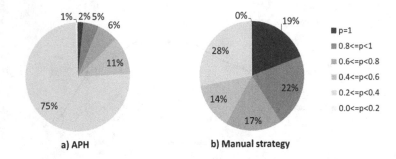

Fig. 5. a) The average block probability distribution of drains over one year using adaptive planning heuristic; b) The average block probability distribution of drains over one year using manual strategy

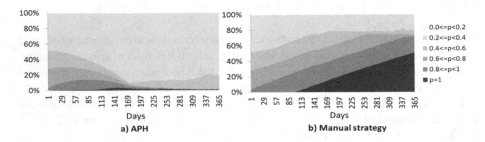

Fig. 6. a) The block probability distribution of drains over one year using adaptive planning heuristic; b) The block probability distribution of drains over one year using manual strategy

because more drains are under a good condition when we apply the APH. This overall increase in correlation comes at a price, the APH is not fast at responding to the calls and damage. On average, the residents' waiting time is about 58 days when the APH is applied, which is about 4.8 times longer than the MS.

Robustness Analysis. Finally, we test robustness of the APH by changing the calling rate γ and the probability of getting damaged. Three scenario are tested here: the first scenario with the total number of calling and damage reports both doubled; the second one with the total number of calling and damage report equal to the Blackpool historic recording data and the third one with the numbers both reduced by half. Sixteen random runs were carried out for each scenario. Table 3 shows the value of fitness is higher under more 'stable' scenario. Furthermore, these results also show that longer 'look ahead' strategy suit more 'stable' scenario. Compared to the strategy without 'look ahead', the best fitnesses improve 0.218%, 0.135% and 0.0029% under scenario one, two, three respectively. 'Look ahead' strategy helps the scheduling process and more effectiveness is obtained under more dynamic situation.

Table 2. The distribution of block probability of drains with different social values

	Overall		Social value $\geq$ 4		Social value $\geq$ 5		Social value $\geq$6	
	Manual	APH	Manual	APH	Manual	APH	Manual	APH
$p = 1$	19.3%	1.8%	14.8%	0.8%	17.9%	0.7%	23%	0.3%
$0.8 \leq p < 1$	21.5%	4.8%	16.1	2.5%	18.9%	2.8%	24.7%	1.5%
$0.6 \leq p < 0.8$	17.3%	6.3%	13.1%	4.1%	14.5%	3.7%	19.3%	2.1%
$0.4 \leq p < 0.6$	13.8%	11.5%	11%	7.3%	11.7%	6.5%	15.2%	3.7%
$0.2 \leq p < 0.4$	**27.8%**	**75%**	**44.6%**	**84.6%**	**36.7%**	**85.7%**	**17.8%**	**91.6%**
$0.0 \leq p < 0.2$	0.2%	0.5%	0.4%	0.7%	0.3%	0.7%	0.1%	0.7%

Table 3. Normalized fitness of the APH under different scenarios

	Best Solution Fitness	Best r	Fitness for r=1	Improvement (%)
First Scenario	100.218	5	100.000	0.2183
Second Scenario	101.809	6	101.672	0.1347
Third Scenario	103.985	7	103.982	0.0029

6 Conclusion

We have studied a dynamic period routing problem with profit (DPVRPP) that
is motivated by the large scale real-life storm drain maintenance problem. Two
difficulties include the large number of drains and the dynamic situation. A
detailed model for the problem has been presented. We have proposed a two
stage method composed of routing and scheduling processes. The main idea is
to effectively and wisely react to the changing drains situation by looking into
the near future. In addition, due to the distribution and large number of drains,
routing them first will help to efficiently use the daily working time and vehicle
resources. Also it helps to decrease the scheduling problem size, which makes
responding to the dynamic situation and replanning every day for large scale
problem possible. The APH approach has been tested on real-world data. Results
show that the APH improves performance by about 58% compared to the current
manual strategy in terms of drains condition by doing preventative maintenance
rather than reactive work. A potential down side is that it then takes longer
to react to (the massively reduced number of) blocked drain reports. We have
shown that looking ahead several days is effective – with the number of days of
look ahead influenced particularly by problem stability.

Acknowledgements. We would like to thank Steve Birdsall of Gaist Solutions
Ltd. for providing data and domain knowledge, as well as the anonymous ref-
erees for their helpful comments. This work is funded by the EPSRC LSCITS
program.

References

1. Christofides, N., Beasley, J.E.: The period routing problem. Networks 14(2), 237–256 (1984)
2. Ball, M.O.: Allocation/routing: Models and algorithms. In: Golden, B.L., Assad, A. (eds.) Vehicle Routing: Methods and Studies, pp. 194–221 (1988)
3. Chao, I.M., Golden, B.L., Wasil, E.: An improved heuristic for the period vehicle routing problem. Networks 26(1), 25–44 (1995)
4. Cordeau, J.F., Gendreau, M., Laporte, G.: A tabu search heuristic for periodic and multi-depot vehicle routing problems. Networks 30(2), 105–119 (1997)
5. Brandão, J., Mercer, A.: A tabu search algorithm for the multi-trip vehicle routing and scheduling problem. Eur. J. Oper. Res. 100(1), 180–191 (1997)
6. Alegre, J., Laguna, M., Pacheco, J.: Optimizing the periodic pick-up of raw materials for a manufacturer of auto parts. Eur. J. Oper. Res. 179(3), 736–746 (2007)
7. Delgado, C., Laguna, M., Pacheco, J.: Minimizing labor requirements in a periodic vehicle loading problem. Comput. Optim. Appl. 32(3), 299–320 (2005)
8. Hemmelmayr, V., Doerner, K., Hartl, R.: A variable neighborhood search heuristic for periodic routing problems. Eur. J. Oper. Res. 195(3), 791–802 (2009)
9. Pirkwieser, S., Raidl, G.R.: Multilevel variable neighborhood search for periodic routing problems. In: Cowling, P., Merz, P. (eds.) EvoCOP 2010. LNCS, vol. 6022, pp. 226–238. Springer, Heidelberg (2010)
10. Baptista, S., Oliveira, R., Zúquete, E.: A period vehicle routing case study. Eur. J. Oper. Res. 139, 220–229 (2002)
11. Goncalves, L., Ochi, L., Martins, S.: A grasp with adaptive memory for a period vehicle routing problem. In: Computational Intelligence for Modelling, Control and Automation, 2005 and International Conference on Intelligent Agents, Web Technologies and Internet Commerce, vol. 1, pp. 721–727 (2005)
12. Angelelli, E., Speranza, M., Savelsbergh, M.: Competitive analysis for dynamic multiperiod uncapacitated routing problems. Networks 49(4), 308–317 (2007)
13. Angelelli, E., Bianchessi, N., Mansini, R., Speranza, M.: Short Term Strategies for a Dynamic Multi-Period Routing Problem. Transportation Research Part C: Emerging Technologies 17(2), 106–119 (2009)
14. Wen, M., Cordeau, J.F., Laporte, G., Larsen, J.: The dynamic multi-period vehicle routing problem. Comput. Oper. Res. 37(9), 1615–1623 (2010)
15. Weibull, W.: A statistical distribution function of wide applicability. Journal of Applied Mechanics 18, 293–297 (1951)
16. Clarke, G., Wright, J.W.: Scheduling of vehicles from a central depot to a number of delivery points. Operations Research 12(4), 568–581 (1964)
17. Mladenović, N., Hansen, P.: Variable neighborhood search. Computers & Operations Research 24, 1097–1100 (1997)
18. Bentley, J.J.: Fast algorithms for geometric traveling salesman problems. ORSA Journal on Computing 4, 387–411 (1992)

Elementary Landscape Decomposition of the Hamiltonian Path Optimization Problem*,**

Darrell Whitley[1] and Francisco Chicano[2]

[1] Dept. of Computer Science, Colorado State University, Fort Collins, CO, USA
whitley@cs.colostate.edu
[2] Dept. de Lenguajes y Ciencias de la Computación, University of Málaga, Spain
chicano@lcc.uma.es

Abstract. There exist local search landscapes where the evaluation function is an eigenfunction of the graph Laplacian that corresponds to the neighborhood structure of the search space. Problems that display this structure are called "Elementary Landscapes" and they have a number of special mathematical properties. The problems that are not elementary landscapes can be decomposed in a sum of elementary ones. This sum is called the *elementary landscape decomposition* of the problem. In this paper, we provide the elementary landscape decomposition for the Hamiltonian Path Optimization Problem under two different neighborhoods.

Keywords: Landscape theory, elementary landscapes, hamiltonian path optimization, quadratic assignment problem.

1   Introduction

Grover [7] originally observed that there exist neighborhoods for Traveling Salesperson Problem (TSP), Graph Coloring, Min-Cut Graph Partitioning, Weight Partitioning, as well as Not-All-Equal-SAT that can be modeled using a wave equation borrowed from mathematical physics. Stadler named this class of problems "elementary landscapes" and showed that if a landscape is elementary, the objective function is an eigenfunction of the Laplacian matrix that describes the connectivity of the neighborhood graph representing the search space. When

---

* This research was sponsored by the Air Force Office of Scientific Research, Air Force Materiel Command, USAF, under grant number FA9550-11-1-0088. The U.S. Government is authorized to reproduce and distribute reprints for Governmental purposes notwithstanding any copyright notation thereon.
** It was also partially funded by the Fulbright program, the Spanish Ministry of Education ("José Castillejo" mobility program), the University of Málaga (Andalucía Tech), the Spanish Ministry of Science and Innovation and FEDER under contract TIN2011-28194 and VSB-Technical University of Ostrava under contract OTRI 8.06/5.47.4142. The authors would also like to thank the organizers and participants of the seminar on Theory of Evolutionary Algorithms (13271) at Schloß Dagstuhl - Leibniz-Zentrum für Informatik.

C. Blum and G. Ochoa (Eds.): EvoCOP 2014, LNCS 8600, pp. 121–132, 2014.
© Springer-Verlag Berlin Heidelberg 2014

the landscape is not elementary it is always possible to write the objective function as a sum of elementary components, called the *elementary landscape decomposition* (ELD) of a problem [6].

Landscape theory has been proven to be quite effective computing summary statistics of the optimization problem. Sutton *et al.* [14] show how to compute statistical moments over *spheres* and *balls* of arbitrary radius around a given solution in polynomial time using the ELD of pseudo-Boolean functions. Chicano and Alba [4] and Sutton and Whitley [13] have shown how the expected value of the fitness of a mutated individual can be exactly computed using the ELD. Measures like the autocorrelation length and the autocorrelation coefficient can be efficiently computed using the ELD of a problem. Chicano and Alba [5] proved that Fitness-Distance Correlation can be exactly computed using landscape theory for pseudo-Boolean functions with one global optimum.

The landscape analysis of combinatorial optimization problems has also inspired the design of new and more efficient search methods. This is the case of the average-constant steepest descent operator for NK-landscapes and MAX-kSAT of Whitley *et al.* [15], the second order partial derivatives of Chen *et al.* [3] and the hyperplane initialization for MAX-kSAT of Hains *et al.* [8].

However, the first step in the landscape analysis is to find the ELD of the problem. In this paper, we present the ELD of the Hamiltonian Path Optimization Problem (HPO) for two different neighborhoods: the reversals and the swaps. This problem has applications in DNA fragment assembling and the construction of radiation hybrid maps. The remainder of the paper is organized as follows. In Section 2 we present a short introduction to landscape theory. Section 3 presents the HPO and its relationship with QAP. In Sections 4 and 5 we present the landscape structre of HPO for the reversals and swaps neighborhood, respectively. Finally, Section 6 concludes the paper and outlines future directions.

2   Background on Landscape Theory

Let (X, N, f) be a landscape, where X is a finite set of candidate solutions, $f : X \to \mathbb{R}$ is a real-valued function defined on X and $N : X \to 2^X$ is the neighborhood operator. The pair (X, N) is called *configuration space* and induces a graph in which X is the set of nodes and an arc between (x, y) exists if $y \in N(x)$. The adjacency and degree matrices of the neighborhood N are:

$$\mathbf{A}_{x,y} = \begin{cases} 1 \text{ if } y \in N(x), \\ 0 \text{ otherwise;} \end{cases} \qquad \mathbf{D}_{x,y} = \begin{cases} |N(x)| \text{ if } x = y, \\ 0 \qquad \text{otherwise.} \end{cases}$$

We restrict our attention to regular neighborhoods, where $|N(x)| = d > 0$ for a constant d, for all $x \in X$. Then, the degree matrix is $\mathbf{D} = d\mathbf{I}$, where $\mathbf{I}$ is the identity matrix. The Laplacian matrix $\mathbf{\Delta}$ associated to the neighborhood is defined by $\mathbf{\Delta} = \mathbf{A} - \mathbf{D}$. In the case of regular neighborhoods it is $\mathbf{\Delta} = \mathbf{A} - d\mathbf{I}$. Any discrete function, f, defined over the set of candidate solutions can be characterized as a vector in $\mathbb{R}^{|X|}$. Any $|X| \times |X|$ matrix can be interpreted as a linear map that acts on vectors in $\mathbb{R}^{|X|}$. For example, the adjacency matrix $\mathbf{A}$

acts on function f as follows $(\mathbf{A}\ f)(x) = \sum_{y \in N(x)} f(y)$, where the component x of $(\mathbf{A}\ f)$ is the sum of the function values of all the neighbors of x. Stadler defines the class of *elementary landscapes* where the function f is an eigenvector (or eigenfunction) of the Laplacian up to an additive constant [11].

Definition 1. *Let (X, N, f) be a landscape and Δ the Laplacian matrix of the configuration space. The landscape is said to be elementary if there exists a constant b, which we call* offset, *and an eigenvalue λ of $-\Delta$ such that $(-\Delta)(f - b) = \lambda(f - b)$. When the neighborhood is clear from the context we also say that f is elementary.*

We use $-\Delta$ instead of Δ in the definition to avoid negative eigenvalues, since Δ is negative semidefinite. In connected neighborhoods, where the graph related to the configuration space (X, N) is connected, the offset b is the average value of the function over the whole search space: $b = \bar{f}$. Taking into account basic results of linear algebra, it can be proven that if f is elementary with eigenvalue λ, $af + b$ is also elementary with the same eigenvalue λ. If f is an elementary function with eigenvalue λ, then the average in the neighborhood of a solution can computed as:

$$\text{Avg}_{y \in N(x)}(f(y)) = f(x) + \frac{\lambda}{d}(\bar{f} - f(x)), \tag{1}$$

known as Grover's wave equation [7], which claims that the average fitness in the neighborhood of a solution x can be computed from the fitness of x. The reader interested in Landscape Theory can refer to the survey by Reidys and Stadler [11].

3   Hamiltonian Path Optimization Problem

Given a complete edge-weighted graph, the Hamiltonian Path Optimization Problem (HPO) consists in finding the minimum-cost path that visits all vertices in the graph exactly once, without visiting any vertex twice. This problem has applications in DNA "Linkage Marker" ordering, and in manufacturing, specifically in "set-up cost" minimization. It is similar to the Traveling Salesperson Problem (TSP), except that the result is not a circuit. The path can start at any vertex in the graph and it can end at any other vertex. There exists a polynomial time tranformation that converts any Hamiltonian Path Optimization Problem over n vertices into a Traveling Salesman Problem over $n + 1$ vertices. Assuming the goal is minimization and that all of the edge weights are positive, an additional *dummy* vertex can be added that is connected to every other vertex with zero cost. The optimal Hamiltonian Path is thus transformed into a circuit by using the dummy vertex (and two zero cost edges) to return to the beginning of the path: the Hamiltonian Path and the Circuit have exactly the same cost. We will return to this observation in the conclusions.

The solutions for the HPO problem can be modeled as permutations, which indicate in which order all the vertices are visited. Let us number all the vertices of the graph and let w be the weight matrix, which we will consider symmetric, where $w_{p,q}$ is the weight of the edge (p,q). The fitness function for HPO is:

$$f(\pi) = \sum_{i=1}^{n-1} w_{\pi(i),\pi(i+1)}, \tag{2}$$

where π is a permutation and $\pi(i)$ is the i-th element in that permutation. HPO is a subtype of a more general problem: the Quadratic Assignment Problem (QAP) [2]. Given two matrices r and w the fitness function of QAP is:

$$f_{QAP}(\pi) = \sum_{i,j=1}^{n} r_{i,j} w_{\pi(i),\pi(j)}. \tag{3}$$

We can observe that (3) generalizes (2) if w is the weight matrix for HPO and we define $r_{i,j} = \delta_{i+1}^{j}$, where δ is the Kronecker delta. HPO has applications in Bioinformatics, in particular in DNA fragment assembling [10] and the construction of radiation hybrid maps [1].

4   Landscape for Reversals

Given a permutation π and two positions i and j with $1 \leq i < j \leq n$, we can form a new permutation π' by reversing the elements between i and j (inclusive). Formally, the new permutation is defined as:

$$\pi'(k) = \begin{cases} \pi(k) & \text{if } k < i \text{ or } k > j, \\ \pi(j+i-k) & \text{if } i \leq k \leq j. \end{cases} \tag{4}$$

Figure 1(a) illustrates the concept of reversal. The *reversal neighborhood* $N_R(\pi)$ of a permutation π contains all the permutations that can be formed by applying reversals to π. Each reversal can be identified by a pair $[i,j]$, which are the starting and ending positions of the reversal. We use square brackets in the reversals to distinguish them from swaps. Then, we have $|N_R(\pi)| = n(n-1)/2$.

In the context of HPO, where we consider a symmetric cost matrix, the solution we obtain after applying the reversal $[i,j] = [1,n]$ to π has the same objective value as π, since it is simply a reversal of the complete permutation. For this reason, we will remove this reversal from the neighborhood. We will call the new neighborhood the *reduced reversal neighborhood*, denoted with N_{RR} to distinguish it from the original reversal neighborhood. We have $|N_{RR}(\pi)| = n(n-1)/2 - 1$. In the remaining of this section we will refer always to the reduced reversal neighborhood unless otherwise stated. The landscape analyzed in this section is composed of the set of permutations of n elements, the reduced reversal neighborhood and the objective function (2). We will use the component model explained in the next section to prove that this landscape is elementary.

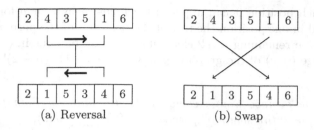

Fig. 1. Examples of reversal and swap for a permutation of size 6

4.1   Component Model

Whitley and Sutton developed a "component" based model that makes it easy to identify elementary landscapes [17]. Let C be a set of "components" of a problem. Each component $c \in C$ has a weight (or cost) denoted by $w(c)$. A solution $x \subseteq C$ is a subset of components and the evaluation function $f(x)$ maps each solution x to the sum of the weights of the components in x: $f(x) = \sum_{c \in x} w(c)$. Finally, let $C - x$ denote the subset of components that do not contribute to the evaluation of solution x. Note that the sum of the weights of the components in $C - x$ is computed by $\sum_{c \in C} w(c) - f(x)$. In the context of the component model, Grover's wave equation can be expressed as:

$$\operatorname*{Avg}_{y \in N(x)}(f(y)) = f(x) - p_1 f(x) + p_2 \left(\sum_{c \in C} w(c) - f(x) \right),$$

where $p_1 = \alpha/d$ is the (sampling) rate at which components that contribute to $f(x)$ are removed from solution x to create a neighboring solution $y \in N(x)$, and $p_2 = \beta/d$ is the rate at which components in the set $C - x$ are sampled to create a neighboring solution $y \in N(x)$. By simple algebra,

$$\operatorname*{Avg}_{y \in N(x)}(f(y)) = f(x) - p_1 f(x) + p_2 \left(\sum_{w \in C} w(c) - f(x) \right) = f(x) + \frac{\lambda}{d}(\bar{f} - f(x)),$$

where $\lambda = \alpha + \beta$, and $\bar{f} = \beta/(\alpha + \beta) \sum_{c \in C} w(c)$ [16].

4.2   Proof of Elementariness

Let us start presenting two lemmas.

Lemma 1. *Given the $n!$ possible Hamiltonian Paths over n vertices, all edges appear the same number of times in the $n!$ solutions and this number is $2(n-1)!$.*

Proof. Each edge (u, v) of the graph can be placed in positions $1, 2, \ldots, n-1$ of the permutation. This means a total of $n-1$ positions where (u, v) can be placed. For each position it can appear as u followed by v or v followed by u. Once it is fixed in one position the rest of the positions of the permutation must be filled with the remaining $n - 2$ elements. This can be done in $(n - 2)!$ ways. Thus, each edge (u, v) of the graph appears in $2 \cdot (n - 1) \cdot (n - 2)! = 2(n - 1)!$ permutations.   □

Lemma 2. *The average fitness value of the Hamiltonian Path Optimization Problem over the entire solution space is:* $\bar{f} = 2/n \sum_{c \in C} w(c)$.

Proof. There are $n(n-1)/2$ edges in the cost matrix, and there are $n-1$ edges in any particular solution. Since all edges uniformly appear in the set of all possible solutions, Lemma 1 implies:

$$\bar{f} = \frac{2(n-1)! \sum_{c \in C} w(c)}{n!} = 2/n \sum_{c \in C} w(c).$$

□

The next theorem presents the main result of this section: it proves that HPO is an elementary landscape for the reduced reversals neighborhood.

Theorem 1. *For the Hamiltonian Path Optimization Problem,*

$$\underset{y \in N_{RR}(x)}{\mathrm{Avg}}(f(y)) = f(x) + \frac{n}{n(n-1)/2 - 1}(\bar{f} - f(x)),$$

and it is an elementary landscape for the reduced reversals neighborhood.

Proof. First, we need to show that all of the $n-1$ edges that contributes to $f(x)$ are uniformly broken by the reverse operator when all of the neighbors of $f(x)$ are generated. The segments to be reversed range in length from 2 to $n - 1$. If the length of the segment is i, then the number of possible segments of length i is $n - (i - 1)$. Let us consider reversals of length i and $n - i + 1$ together, where $1 < i \leq n/2$. The reversal of length i will break the first and last $i - 1$ edges in the permutation only once, but it will break all interior edges twice. However the reversal of length $n - i + 1$ will only break the first and last $i - 1$ edges, and it will break these edges only once. Thus, grouping these together, all edges are broken twice for each value of i.

When n is even, the reversals of length $n/2$ and $n/2 + 1$ are grouped together and the pairing is complete. Thus, for $i = 2$ to $n/2$ all edges are broken twice, and thus every edge is broken $2(n/2 - 1) = (n - 2)$ times and $\alpha = n - 2$.

When n is odd, reversals of length $(n + 1)/2$ are a special case. For $i = 2$ to $(n - 1)/2$ all edges are broken twice, so all edges have been broken $2((n-1)/2 - 1) = (n - 3)$ times, but this does not count the reversal of length $(n + 1)/2$. When the reversal of length $(n + 1)/2$ is applied (and is not paired with another reversal) each edge is broken exactly once. Thus all edges are now broken $(n - 2)$ times and $\alpha = n - 2$.

Next we show that all weights in the cost matrix in the set $C - x$ are uniformly sampled by the reverse operator. Consider the vertex v_i in the permutation. Holding i fixed for the moment, consider all cuts that are made at location i and all feasible locations $j < i$. When all possible values of j are considered, this causes all of the vertices in the permutation left of vertex v_i to come into a position adjacent to v_i except for v_{i-1} which is already adjacent. Next consider a cut at location $i+1$ (i is still fixed) and all feasible locations $m > i+1$. When all of the possible of value of m are considered all of the vertices in the permutation to the right of vertex v_i are moved into a position adjacent to v_i except v_{i+1}. Thus, in these cases v_i does not move, but every edge not in the solution x that is incident on vertex i is sampled once. Since this is true for all vertices, it follows that every edge not in the current solution is sampled twice ($\beta = 2$): once for each of the vertices in which it is incident. Therefore, summing over all the neighbors: $d \cdot \text{Avg}(f(y))_{y \in N(x)} = d \cdot f(x) - (n-2)f(x) + 2 \left(\sum_{c \in C} w(c) - f(x) \right).$

Computing the average over the neighborhood and taking into account the result of Lemma 2:

$$\text{Avg}_{y \in N(x)}(f(y)) = f(x) - \frac{n-2}{n(n-1)/2 - 1} f(x) + \frac{2}{n(n-1)/2 - 1} \left(\sum_{c \in C} w(c) - f(x) \right)$$

$$= f(x) + \frac{n}{n(n-1)/2 - 1} (\bar{f} - f(x)).$$

$\square$

5   Landscape Structure for Swaps

Given a permutation π, we can build a new permutation π' by swapping two positions i and j in the permutation. The new permutation is defined as:

$$\pi'(k) = \begin{cases} \pi(k) & \text{if } k \neq i \text{ and } k \neq j, \\ \pi(i) & \text{if } k = j, \\ \pi(j) & \text{if } k = i. \end{cases} \tag{5}$$

Figure 1(b) illustrates the concept of swap. The *swap neighborhood* $N_S(\pi)$ of a permutation π contains all the permutations that can be formed by applying swaps to π. Each swap can be identified by the pair (i, j) of positions to swap. The cardinality of the swap neighborhood is $|N_S(\pi)| = n(n-1)/2$. Unless otherwise stated, we will refer always to the swap neighborhood in this section.

We will analyze the landscape composed of the set of permutations of n elements, the swap neighborhood and the objective function (2). The approach used for this analysis will be different than the one in the previous section. Instead of using the component model we will base our results in the analysis of the QAP landscape done in [6]. The reasons for not using the component model will be discussed at the end of the section.

5.1   Previous Results for QAP

According to Chicano *et al.* [6], the ELD of the QAP for the swap neighborhood is composed of three components. The objective function analyzed in [6] is more general than (3), it corresponds to the Lawler version of QAP [9]:

$$f(\pi) = \sum_{i,j,p,q=1}^{n} \psi_{i,j,p,q} \delta_{\pi(i)}^{p} \delta_{\pi(j)}^{q}, \qquad (6)$$

where δ is the Kronecker delta. The correspondence with (3) is $\psi_{i,j,p,q} = r_{i,j} w_{p,q}$. The elementary landscape decomposition of the Lawler QAP is given next.

Theorem 2 (from [6]). *The landscape composed of the permutations of n elements, the swap neighborhood and the objective function (6) can be decomposed as the sum of at most three elementary landscapes with eigenvalues $\lambda_1 = n$, $\lambda_2 = 2n$ and $\lambda_3 = 2(n-1)$. The definition of these elementary components are:*

$$f_n(\pi) = \frac{1}{n(n-2)} \sum_{\substack{i,j=1 \\ i \neq j}}^{n} \sum_{\substack{p,q=1 \\ p \neq q}}^{n} \psi_{i,j,p,q} \, \phi_{(i,j),(p,q)}^{n}(\pi) + \sum_{i,p=1}^{n} \psi_{i,i,p,p} \, \delta_{\pi(i)}^{p}, \quad (7)$$

$$f_{2n}(\pi) = \frac{1}{2n} \sum_{\substack{i,j=1 \\ i \neq j}}^{n} \sum_{\substack{p,q=1 \\ p \neq q}}^{n} \psi_{i,j,p,q} \, \phi_{(i,j),(p,q)}^{2n}(\pi), \qquad (8)$$

$$f_{2(n-1)}(\pi) = \frac{1}{2(n-2)} \sum_{\substack{i,j=1 \\ i \neq j}}^{n} \sum_{\substack{p,q=1 \\ p \neq q}}^{n} \psi_{i,j,p,q} \, \phi_{(i,j),(p,q)}^{2(n-1)}(\pi), \qquad (9)$$

where the ϕ functions are defined using as base the $\phi_{(i,j),(p,q)}^{\alpha,\beta,\gamma,\varepsilon,\zeta}$ function as $\phi_{(i,j),(p,q)}^{n} = \phi_{(i,j),(p,q)}^{n-1,3-n,0,2-n,1-n}$, $\phi_{(i,j),(p,q)}^{2n} = \phi_{(i,j),(p,q)}^{n-1,3-n,0,2,1}$ and $\phi_{(i,j),(p,q)}^{2(n-1)} = \phi_{(i,j),(p,q)}^{n-3,n-3,0,0,1}$. The $\phi_{(i,j),(p,q)}^{\alpha,\beta,\gamma,\varepsilon,\zeta}$ function is defined as:

$$\phi_{(i,j),(p,q)}^{\alpha,\beta,\gamma,\varepsilon,\zeta}(\pi) = \begin{cases} \alpha \text{ if } \pi(i) = p \wedge \pi(j) = q, \\ \beta \text{ if } \pi(i) = q \wedge \pi(j) = p, \\ \gamma \text{ if } \pi(i) = p \oplus \pi(j) = q, \\ \varepsilon \text{ if } \pi(i) = q \oplus \pi(j) = p, \\ \zeta \text{ if } \pi(i) \notin \{p,q\} \wedge \pi(j) \notin \{p,q\}, \end{cases} \qquad (10)$$

where $\oplus$ denotes the exclusive OR. The function f can be written in a compact form as: $f = f_n + f_{2n} + f_{2(n-1)}$.

In the next subsection we analyze this decomposition, providing some properties that are useful to find the elementary landscape decomposition of the HPO under the swap neighborhood.

5.2   Elementary Landscape Decomposition of the HPO

The functions $\phi_{(i,j),(p,q)}^{n}$, $\phi_{(i,j),(p,q)}^{2n}$, and $\phi_{(i,j),(p,q)}^{2(n-1)}$ defined above have some properties that make f_{2n} and $f_{2(n-1)}$ vanish when the matrices r and w fulfill some concrete conditions. The next proposition summarizes these properties:

Proposition 1. *The functions $\phi^n_{(i,j),(p,q)}$, $\phi^{2n}_{(i,j),(p,q)}$, and $\phi^{2(n-1)}_{(i,j),(p,q)}$ defined in Theorem 2 hold the next equalities:*

$$\phi^\lambda_{(i,j),(p,q)} = \phi^\lambda_{(j,i),(q,p)} \quad \text{for } \lambda = n, 2n, 2(n-1), \tag{11}$$

$$\phi^{2n}_{(i,j),(p,q)} + \phi^{2n}_{(j,i),(p,q)} = 2, \tag{12}$$

$$\phi^{2(n-1)}_{(i,j),(p,q)} = \phi^{2(n-1)}_{(j,i),(p,q)}. \tag{13}$$

Proof. The first equation (11) follows from the fact that exchanging i and j at the same time as p and q in any of the branch conditions of the function (10) leaves the condition unchanged, thus the function is the same.

In order to prove Eqs. (12) and (13) we can observe that exchanging i and j in (10) is equivalent to swapping the values of α and β (first and second branches) and those of γ and ε (third and fourth branches). The fifth branch is left unchanged. Thus, Eq. (13) is a direct consequence of this swap of branch values and (12) can be proven as follows:

$$\phi^{2n}_{(i,j),(p,q)}(\pi) + \phi^{2n}_{(j,i),(p,q)}(\pi)$$

$$= \begin{cases} n-1 \\ 3-n \\ 0 \\ 2 \\ 1 \end{cases} + \begin{cases} 3-n \\ n-1 \\ 2 \\ 0 \\ 1 \end{cases} = \begin{cases} 2 \text{ if } \pi(i) = p \wedge \pi(j) = q, \\ 2 \text{ if } \pi(i) = q \wedge \pi(j) = p, \\ 2 \text{ if } \pi(i) = p \oplus \pi(j) = q, \\ 2 \text{ if } \pi(i) = q \oplus \pi(j) = p, \\ 2 \text{ if } \pi(i) \notin \{p,q\} \wedge \pi(j) \notin \{p,q\}. \end{cases}$$

$\square$

Let us now analyze the consequences of these properties for the elementary landscape decomposition of a QAP instance.

Theorem 3. *Let us consider the elementary landscape decomposition of (6) for the swap neighborhood given in Theorem 2. Then we have:*

- *If any of the matrices r or w is symmetric then f_{2n} is constant.*
- *If any of the matrices r or w is antisymmetric then $f_{2(n-1)}$ is zero.*

Proof. We will prove the theorem assuming that r has the required property (symmetric or antisymmetric) and then we will prove that the corresponding elementary components are still constant or zero if we exchange r and w.

Let us assume that matrix r is symmetric ($r_{i,j} = r_{j,i}$), then we can write:

$$f_{2n} = \frac{1}{2n} \sum_{\substack{i,j=1 \\ i<j}}^{n} \sum_{\substack{p,q=1 \\ p\neq q}}^{n} r_{i,j} w_{p,q} \left(\phi^{2n}_{(i,j),(p,q)} + \phi^{2n}_{(j,i),(p,q)} \right) \quad \text{by (12)}$$

$$= \frac{1}{2n} \sum_{\substack{i,j=1 \\ i<j}}^{n} \sum_{\substack{p,q=1 \\ p\neq q}}^{n} 2 r_{i,j} w_{p,q} = \frac{1}{n} \left(\sum_{\substack{i,j=1 \\ i<j}}^{n} r_{i,j} \right) \left(\sum_{\substack{p,q=1 \\ p\neq q}}^{n} w_{p,q} \right), \tag{14}$$

which is a constant value that only depends on the instance, not the solution. Let us now assume that r is antisymmetric $(r_{i,j} = -r_{j,i})$, then we can write:

$$f_{2(n-1)} = \frac{1}{2(n-2)} \sum_{\substack{i,j=1 \\ i<j}}^{n} \sum_{\substack{p,q=1 \\ p \neq q}}^{n} r_{i,j} w_{p,q} \left(\phi_{(i,j),(p,q)}^{2(n-1)} - \phi_{(j,i),(p,q)}^{2(n-1)} \right) = 0.$$

Let us call $g(\pi)$ to the fitness function we obtain by exchanging matrices r and w in the definition of the fitness function (6) and let us call $f(\pi)$ to the original fitness function. The relationship between these two functions is as follows:

$$g(\pi) = \sum_{i,j=1}^{n} \sum_{p,q=1}^{n} w_{i,j} r_{p,q} \delta_{\pi(i)}^{p} \delta_{\pi(j)}^{q} \quad \text{renaming indices } (p \leftrightarrow i \text{ and } q \leftrightarrow j)$$

$$= \sum_{p,q=1}^{n} \sum_{i,j=1}^{n} w_{p,q} r_{i,j} \delta_{\pi(p)}^{i} \delta_{\pi(q)}^{j} = \sum_{i,j=1}^{n} \sum_{p,q=1}^{n} r_{i,j} w_{p,q} \delta_{\pi^{-1}(i)}^{p} \delta_{\pi^{-1}(j)}^{q} = f(\pi^{-1}).$$

The elementary landscape decomposition of $g(\pi) = g_n(\pi) + g_{2n}(\pi) + g_{2(n-1)}(\pi)$ can thus be written based on the elementary landscape decomposition of $f(\pi)$ as: $g_\lambda(\pi) = f_\lambda(\pi^{-1})$ for $\lambda = n, 2n, 2(n-1)$. If the w matrix of the original fitness function f is symmetric or antisymmetric then the r matrix of the g function will have the same property, and, as we have proven above, $g_{2n} = constant$ or $g_{2(n-1)} = 0$, respectively. But this means that also $f_{2n} = constant$ or $f_{2(n-1)} = 0$, respectively. $\qquad \square$

Since the addition of constant values does not affect the number of elementary components of a landscape, when one of the components is constant in the previous theorem, this component vanishes from the elementary landscape decomposition. In other words, the number of elementary components is reduced whenever any of the conditions in the previous theorem holds. For example, an instance of QAP having a symmetric w will have at most two elementary components, instead of three. The conditions can co-occur in the same instance. If an instance of QAP has one of the matrices symmetric and the other one antisymmetric it will be an elementary landscape, since f_{2n} and $f_{2(n-1)}$ are constants. For HPO we have the next result.

Corollary 1. *The Hamiltonian Path Optimization Problem with symmetric cost matrix is the sum of at most two elementary components in the swap neighborhood. One possible expression for these two components is:*

$$f_n = \frac{1}{n(n-2)} \sum_{i=1}^{n-1} \sum_{\substack{p,q=1 \\ p \neq q}}^{n} w_{p,q} \phi_{(i,i+1),(p,q)}^{n} + \frac{n-1}{n} \sum_{\substack{p,q=1 \\ p<q}}^{n} w_{p,q}, \qquad (15)$$

$$f_{2(n-1)} = \frac{1}{2(n-2)} \sum_{i=1}^{n-1} \sum_{\substack{p,q=1 \\ p \neq q}}^{n} w_{p,q} \phi_{(i,i+1),(p,q)}^{2(n-1)}, \qquad (16)$$

where f_n, and $f_{2(n-1)}$ are elementary with eigenvalues n and $2(n-1)$, respectively, and the objective function of HPO f can be written as: $f = f_n + f_{2(n-1)}$.

Proof. Since the matrix w is symmetric in the HPO, according to Theorem 3 the f_{2n} component is constant and its value is: $f_{2n} = \frac{n-1}{n} \sum_{\substack{p,q=1 \\ p<q}} w_{p,q}$, where we use (14) and take into account that the r matrix contains $n-1$ ones in the upper triangle. We sum this constant to the f_n component (without changing its elementariness) and we have the claimed result after some algebraic simplifications in the equations.   □

The landscape of HPO with the swap neighborhood is the sum of two elementary landscapes. This is the main reason why the component model is not useful for this analysis. The component model can only be used to prove that a landscape is elementary but it is not able to describe landscapes that are sums of more than one elementary component. Furthermore, it is not able to prove the elementariness of some elementary landscapes. This happens, in particular, in elementary landscapes with irrational eigenvalues.

Another mathematical tool to provide the elementary landscape decomposition of optimization problems is the spectral theory of quasi-Abelian Cayley graphs. When the solution set is an Abelian group this theory provides an easy and elegant elementary landscape decomposition of any objective function. This happens for example, in the case of the q-ary hypercube [12]. When the solution set is not an Abelian group, as in the case of the permutations, the math involved is more complex but still give elegant answers if the set of generators that define the neighborhood is the union of conjugacy classes of the group [11]. The generators of the swap neighborhood (2-cycles) form a conjugacy class, so the spectral theory could be applied in this case; but the generators of the reversal (and reduced reversal) neighborhood are not the union of conjugacy classes, what makes the component model to be, as far as we know, the best methodology to prove the elementariness of the landscape.

6   Conclusions and Future Work

In this work we have proven that the Hamiltonian Path Optimization Problem is elementary under the reversals neighborhood and is the sum of two elementary components in the case of the swaps neighborhood. We have also provided the expressions for the elementary components in the latter case.

In the two neighborhoods considered in the paper the number of moves is $\Theta(n^2)$. For problems that are extremely large, e.g., with billions of vertices, it may not be reasonable to look at all possible moves. On the other hand, some form of heuristic might be used to decide if there exists a smaller subset of locations where the permutation might be reasonably be broken, while other edges are not broken. In this way, a problem with billions of vertices might be mapped onto a problem with only hundreds of thousands vertices. This leads to the so-called Partial Neighborhoods [17]. An interesting question to analyze is the landscape structure for these partial neighborhoods.

At the beginning of Section 3, we noted that every Hamiltonian Path Optimization Problem over n vertices can be converted into a Traveling Salesman Problems over $n + 1$ vertices. In practice, this means that existing TSP solvers can be use to solve the Hamiltonian Path Optimization Problem. One interesting question is whether the elementary landscapes which exist for the TSP can be mapped onto the elementary landscapes which we have defined for the Hamiltonian Path Optimization Problem.

References

1. Agarwala, R., Applegate, D.L., Maglott, D., Schuler, G.D.: A fast and scalable radiation hybrid map construction and integration strategy. Genome Research 10(3), 350–364 (2000)
2. Burkard, R.E.: Quadratic Assignment Problems. In: Handbook of Combinatorial Optimization, 2nd edn., pp. 2741–2815. Springer (2013)
3. Chen, W., Whitley, D., Hains, D., Howe, A.: Second order partial derivatives for NK-landscapes. In: Proceeding of GECCO, pp. 503–510. ACM (2013)
4. Chicano, F., Alba, E.: Exact computation of the expectation curves of the bit-flip mutation using landscapes theory. In: Proc. of GECCO, pp. 2027–2034 (2011)
5. Chicano, F., Alba, E.: Exact computation of the fitness-distance correlation for pseudoboolean functions with one global optimum. In: Hao, J.-K., Middendorf, M. (eds.) EvoCOP 2012. LNCS, vol. 7245, pp. 111–123. Springer, Heidelberg (2012)
6. Chicano, F., Whitley, L.D., Alba, E.: A methodology to find the elementary landscape decomposition of combinatorial optimization problems. Evolutionary Computation 19(4), 597–637 (2011)
7. Grover, L.K.: Local search and the local structure of NP-complete problems. Operations Research Letters 12, 235–243 (1992)
8. Hains, D., Whitley, D., Howe, A., Chen, W.: Hyperplane initialized local search for MAXSAT. In: Proceeding of GECCO, pp. 805–812. ACM (2013)
9. Lawler, E.L.: The quadratic assignment problem. Manage. Sci. 9, 586–599 (1963)
10. Parsons, R., Forrest, S., Burks, C.: Genetic algorithms, operators, and DNA fragment assembly. Machine Learning 21, 11–33 (1995)
11. Reidys, C.M., Stadler, P.F.: Combinatorial landscapes. SIAM Review 44(1), 3–54 (2002)
12. Sutton, A.M., Chicano, F., Whitley, L.D.: Fitness function distributions over generalized search neighborhoods in the q-ary hypercube. Evol. Comput. 21(4) (2013)
13. Sutton, A.M., Whitley, D., Howe, A.E.: Mutation rates of the (1+1)-EA on pseudoboolean functions of bounded epistasis. In: Proc. of GECCO, pp. 973–980 (2011)
14. Sutton, A.M., Whitley, L.D., Howe, A.E.: Computing the moments of k-bounded pseudo-boolean functions over hamming spheres of arbitrary radius in polynomial time. Theoretical Computer Science 425, 58–74 (2011)
15. Whitley, D., Chen, W.: Constant time steepest descent local search with lookahead for NK-landscapes and MAX-kSAT. In: Proc. of GECCO, pp. 1357–1364 (2012)
16. Whitley, D., Sutton, A.M., Howe, A.E.: Understanding elementary landscapes. In: Proc. of GECCO, pp. 585–592 (2008)
17. Whitley, L.D., Sutton, A.M.: Partial neighborhoods of elementary landscapes. In: Proc. of GECCO, pp. 381–388 (2009)

Gaussian Based Particle Swarm Optimisation and Statistical Clustering for Feature Selection

Mitchell C. Lane[1], Bing Xue[1], Ivy Liu[2], and Mengjie Zhang[1]

[1] School of Engineering and Computer Science
[2] School of Mathematics, Statistics and Operations Research,
Victoria University of Wellington, P.O. Box 600, Wellington 6140, New Zealand
{Mitchell.Lane,Bing.Xue,Mengjie.Zhang}@ecs.vuw.ac.nz,
Ivy.Liu@msor.vuw.ac.nz

Abstract. Feature selection is an important but difficult task in classification, which aims to reduce the number of features and maintain or even increase the classification accuracy. This paper proposes a new particle swarm optimisation (PSO) algorithm using statistical clustering information to solve feature selection problems. Based on Gaussian distribution, a new updating mechanism is developed to allow the use of the clustering information during the evolutionary process of PSO based on which a new algorithm (GPSO) is developed. The proposed algorithm is examined and compared with two traditional algorithms and a PSO based algorithm which does not use clustering information on eight benchmark datasets of varying difficulty. The results show that GPSO can be successfully used for feature selection to reduce the number of features and achieve similar or even better classification performance than using all features. Meanwhile, it achieves better performance than the two traditional feature selection algorithms. It maintains the classification performance achieved by the standard PSO for feature selection algorithm, but significantly reduces the number of features and the computational cost.

Keywords: Particle swarm optimisation, Gaussian distribution, Statistical clustering, Feature selection.

1 Introduction

Feature selection is a process of selecting a small subset of relevant features from the original large feature set, which can reduce the dimensionality of the data and increase the performance of a machine learning technique (e.g. a classification algorithm). It becomes increasingly important in data mining and machine learning because of the advances of data collection techniques, which increases the total number of features included in the dataset. Existing feature selection algorithms can be broadly classified into two categories: filter and wrapper approaches [1]. Filter approaches are independent of any learning algorithm while wrapper approaches include a classification/learning algorithm as part of the evaluation function. Therefore, wrapper approaches can often achieve better accuracy than filter approaches [1].

Feature selection is a challenging task, which has a large search space with 2^n possible points, where n is the total number of features in the dataset. This leads to the

C. Blum and G. Ochoa (Eds.): EvoCOP 2014, LNCS 8600, pp. 133–144, 2014.
© Springer-Verlag Berlin Heidelberg 2014

problems of the high computational cost and stagnation in local optima in most existing feature selection approaches. Particle swarm optimisation (PSO) [2, 3] is a powerful global search technique, which is computationally less expensive than other evolutionary computation techniques such as genetic programming (GP) and genetic algorithms (GAs) [4]. Therefore, PSO has been successfully applied to many areas, including feature selection [5–7].

Feature interaction is a common and complex problem in classification [1], which also makes feature selection a hard problem. Feature interaction may change the relationship between a feature(s) and the class labels. Due to feature interaction, an individually relevant feature may become redundant and a weakly relevant feature may become highly useful when combining with other features. The removal or addition of some features needs to consider the appearance or absence of other features. Therefore, the optimal feature subset is a group of complementary features that working together can increase the classification performance.

Many statistical measures have been applied to form the evaluation function in feature selection algorithms [8]. However, all of them are used in filter approaches. Statistical clustering methods [9, 10] can group relatively homogeneous features together based on a statistical model. This method considers all features simultaneously and takes feature interaction into account. Features in the same cluster are similar and they are dissimilar to features in other clusters. Since feature interaction is an important factor in feature selection, the statistical feature interaction information found by the clustering method can be used to develop a good feature selection algorithm. However, this has not been seriously investigated to date.

1.1 Goals

The overall goal of this paper is to investigate the use of statistical clustering information in PSO for feature selection. To achieve this goal, a statistical clustering method is performed as a preprocessing step on part of the training set to group features into different clusters. A Gaussian based updating mechanism is developed to incorporate the clustering information during the evolutionary process of PSO. A new PSO based feature selection algorithm named GPSO is then developed to reduce the number of features and increase the classification accuracy. Specifically, we will investigate:

- whether GPSO with the developed Gaussian updating mechanism can successfully utilise the clustering information to select a small subset of features to achieve similar or even better classification performance than using all features;
- whether GPSO can achieve better performance than the standard PSO for feature selection without clustering information, and
- whether GPSO can outperform two traditional feature selection algorithms.

2 Background

2.1 Particle Swarm Optimisation (PSO)

PSO is an evolutionary computation (EC) technique, which imitates the social behaviours of birds flocking and fish schooling [2, 3]. PSO uses a swarm of particles to search for the optimal solution, where each particle represents a possible solution in the search space. Each particle has a position vector, $x_i = (x_{i1}, x_{i2}, ..., x_{iD})$, and a velocity vector, $v_i = (v_{i1}, v_{i2}, ..., v_{iD})$, where D is the dimensionality. During the evolutionary process, each particle remembers its previous best position (*pbest*) and the best position found so far by the swarm (*gbest*). In binary PSO (BPSO)[11], each element in the position vector is a binary value. The velocity represents the probability of an element in the position taking value 1. To achieve this, a sigmoid function $s(v_{id})$ is used to transform v_{id} to $(0, 1)$. BPSO updates the position and velocity of each particle according to Equations 1 and 2.

$$v_{id}^{t+1} = w * v_{id}^t + c_1 * r_{1i} * (p_{id} - x_{id}^t) + c_2 * r_{2i} * (p_{gd} - x_{id}^t) \qquad (1)$$

$$x_{id} = \begin{cases} 1, & \text{if } rand() < s(v_{id}) \\ 0, & otherwise \end{cases} \qquad (2)$$

where

$$s(v_{id}) = \frac{1}{1 + e^{-v_{id}}} \qquad (3)$$

where t denotes the tth iteration in the search process. d denotes the dth dimension in the search space. w is the inertia weight. c_1 and c_2 are acceleration constants. r_{1i}, r_{2i} and $rand()$ are random values uniformly distributed in $[0, 1]$. p_{id} and p_{gd} represent the value of *pbest* and *gbest* in the dth dimension, respectively. v_{id}^t is limited by a predefined maximum velocity v_{max}, where $v_{id}^t \in [-v_{max}, v_{max}]$.

When using BPSO for feature selection, the dimensionality of the search space is the total number of features in the dataset. "1" in the position vector means the corresponding feature is selected and "0" otherwise [5].

2.2   Related Work on Feature Selection

A number of feature selection algorithms have been proposed, which can be seen in [1, 5, 12]. Due to the page limit, only typical EC based feature selection algorithms and the role of statistics are reviewed here.

EC Approaches for Feature Selection. Zhu et al. [13] proposed a feature selection method using a memetic algorithm that is a combination of local search and GA. Experiments show that this algorithm outperforms GA alone and other algorithms. Neshatian et al. [14] proposed a feature ranking method for feature selection, where each feature is assigned a score according to the frequency of its appearance in a collection of GP trees and the fitness of those trees. Feature selection can be achieved by using the top-ranked features for classification. Based on ant colony optimisation (ACO), Kanan and Faez [15] developed a wrapper feature selection algorithm, which outperforms GA and other ACO based algorithms on a face detection dataset, but its performance has not been tested on other problems. He et al. [16] applied a binary differential evolution (BDE) algorithm to filter feature selection with a mutual information based fitness function. However, the proposed algorithm is not compared with any other algorithm and the datasets used include a relatively small number (maximum 56) of features. Al-Ani et al. [17] also proposed a DE based method, where features are distributed to a set of wheels and DE is employed to select features from each wheel. This algorithm can significantly reduce the number of features and improve the classification performance.

Chuang et al. [5] proposed a PSO based algorithm that resets *gbest* if it maintains the same value after several iterations. The experiments on cancer-related gene expression datasets show that the proposed algorithm can select a small number of features to improve the classification performance. Xue et al. [18] developed new initialisation and *pbest* and *gbest* updating mechanisms in PSO for feature selection, which can increase the classification accuracy and reduce both the number of features and the computational time. Wang et al. [19] redefined the velocity in BPSO as the number of elements that should be changed in the position. The experiments show that the proposed approach is computationally less expensive than GA. Fdhila et al. [20] applied a multi-swarm PSO algorithm to solve feature selection problems. However, the computational cost of the proposed algorithm is high because it involves parallel evolutionary processes and multiple sub-swarms with a relative large number of particles. Yang et al. [21] proposed two PSO based feature selection approaches based on two inertia weight setting methods. The results show that the two algorithms can outperform sequential forward search, sequential forward floating search, sequential GA and different hybrid GAs. Xue et al. [12, 22] also proposed a PSO based multi-objective approach for feature selection, which shows that the PSO based approach outperforms three other commonly used EC based multi-objective algorithms, i.e. NSGAII, SPEA2, and PAES.

Javani et al. [23] applied PSO for feature selection and clustering in machine learning, where each particle is used to optimise the weights for all features and cluster center values. feature selection is achieved by omitting features with a low weight. However, features with a low weight may be useful because of feature interaction and the removal may reduce the performance of the feature subset. **Note** that the clustering problem here is a machine learning task which aims to group **instances** into different clusters. This is different from the statistical clustering used in this paper, which aims to group **features** into different clusters.

Statistics in Feature Selection. Many statistical methods can be used to reduce the dimensionality of a dataset [8], such as principal component analysis, linear discriminant analysis, or canonical correlation analysis. However, most of them are not feature selection approaches because they create new features. Some researchers introduce statistical measures to evaluate the relationship between a feature and the class labels, which are then used in feature selection to evaluate the goodness of the selected features. Based on a statistical discrepancy measure, Jakub Segen [24] developed a feature selection method, which starts with the feature that best distinguishes the classes, and iteratively adds features which in combination with the chosen features improve the classification discrimination. Relief [25] uses a statistical method to select the relevant features, where each feature has a score indicating its relevance to the class labels. Relief selects all the relevant features. However, the selected features may still have redundancy because Relief does not consider the redundancy between the relevant features. Many other statistical measures such as Pearson's correlation and least square regression error, have been used in feature selection to score the significance of features in class separability.

Clustering analysis is an important class of statistical techniques that can be applied to group features/variables to a number of clusters. A statistical clustering method can group relatively homogeneous features together taking feature interactions into account [9, 10]. A statistical clustering method usually considers feature interaction in the dataset. Therefore, the statistical feature interaction information found by a statistical clustering method can be used to develop a good feature selection algorithm, but this has not been seriously investigated.

Based on PSO and a statistical clustering method [9, 10] that groups features into different clusters and similar features to the same cluster, Lane et al. [6] proposed a feature selection algorithm, which uses PSO to select one feature from each cluster. The results show that by selecting a representative feature from each cluster, the proposed algorithm can significantly reduce the number of features and increase the classification performance. This shows the the statistical clustering information (i.e. feature clusters) can provide useful information in feature selection. Therefore, this work will also utilise such information to further develop the new approach.

3   The Proposed Approach

We use a newly developed clustering method based on statistical models proposed by Pledger and Arnold [9] and Matechou et. al. [10] to group features into different clusters. Due to the page limit, it is not described here. The statistical clustering method is performed as a preprocessing step on a small number of training instances to group features into different clusters.

Features in the same cluster are considered as similar features. Therefore, to use statistical clustering information for feature selection, on one hand, a single feature can be selected as a representative of its associated cluster. On the other hand, features from the same cluster might still be complementary to each other, which means that multiple features may be needed from a single feature cluster. Therefore, we want to consider feature clustering and feature interaction information to develop a new PSO approach to selecting features based on the obtained feature clusters, which is different from the traditional PSO based approach that selects features based on the whole feature set. The new approach is expected to encourage the selection of a *single* feature from each cluster, but when needed, it can also select multiple features from the same cluster. However, the original updating mechanism in PSO does not consider clustering information. Therefore, a new position updating mechanism is needed.

In PSO for feature selection, the position of a particle represents one feature subset, but the traditional position updating mechanism PSO does not consider the clustering information. Based on a Gaussian distribution (i.e. normal distribution) function, the new position updating mechanism is proposed to consider the clustering information, which first determines the number of features that will be selected from a cluster, and then determines the selection of individual features from that cluster.

3.1   Determine the Number of Features Selected

Since a small number of features is preferred, there should be a relatively large (small) probability to select a small (large) number of features from a given feature cluster.

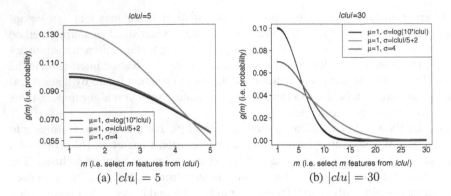

Fig. 1. The effects of the standard deviation functions upon two Gaussian distributions (colour)

Gaussian distribution is used here to determine the probability of selecting a certain number (m) of features. Gaussian distribution is typically shown by $N(\mu, \sigma)$, where μ is the mean and σ is the standard deviation. The output of the Gaussian function is used here as the probability of selecting m of features from a cluster. In Gaussian function, the output value is the largest when using μ as input. Since selecting only 1 feature from each cluster is the ideal case, which should have the largest probability, $\mu = 1$ is used here. σ determining the change of the probability is a key factor, which should be defined according to the feature cluster size, i.e. the number of features in this cluster. A logarithmic function using the cluster size ($|clu|$) as the input variable, $\sigma = log(10 \times |clu|)$, is used to determine σ.

Based on $\mu = 1$ and $\sigma = log(10 \times |clu|)$, the Gaussian distribution function is built to calculate the probability of selecting m ($1 \leq m \leq |clu|$) features from a given cluster, which is shown by Equation 4.

$$g(m) = \frac{\exp(-\frac{(m-1)^2}{2log^2(10 \times |clu|)})}{\sqrt{2\pi}\log(10 \times |clu|)} \tag{4}$$

Fig. 1 plots the Gaussian function shown by Equation 4, where $|clu| = 5$ in Fig. 1(a) is used as a representative of a small feature cluster, and $|clu| = 30$ in Fig. 1(b) is used as a representative of a large feature cluster. Fig. 1 also plots the Gaussian distribution of using a constant $\sigma = 4$ and and a linearly changing σ ($\sigma = |clu|/5 + 2$), which are used for comparison purposes to explain why $\sigma = log(10 \times |clu|)$ is chosen here. From Fig. 1(a), it can be seen that $\sigma = log(10 \times |clu|)$ provides a chance of selecting 1, 2, 3 or 4 features that is more even than the linear function, which favors selecting 1 or 2 features from the small cluster. From Fig. 1(b), it can be observed that $\sigma = log(10 \times |clu|)$ provides a much smaller chance for selecting more than 10 features than the other two standard deviation functions. Therefore, fewer redundant features will be introduced when using $\sigma = log(10 \times |clu|)$ since features are similar within a cluster.

For a given feature cluster, a desired feature list is formed by adding the features if a random value is smaller than $s(v_{id})$. If there are $|DF|$ features, the sum of all the possible $g(m)$ values should be 1. Therefore, $g(m)$ is normalised to make sure

$\sum_{m=1}^{|DF|} g(m) = 1$. Based on the normalised $g(m)$ values, a "roulette wheel selection" is performed here to determine the value of m. Note that the "roulette wheel selection" is performed on features within a cluster (not on individuals within a swarm/population). It is used here to ensure that the large $g(m)$ will have a large chance to be selected, but the small $g(m)$ will also have a chance to be selected (not completely ruled out).

3.2 How to Select Features

When using PSO for feature selection, each feature corresponds to one dimension in the position and velocity. "1" in the position means the corresponding feature is selected. Selecting m features from a cluster means m dimensions in the position are updated to "1" and all other dimensions in the same cluster are updated to "0".

In the proposed algorithm, m features are chosen based on the maximum probability mechanism, where the motivation is that the velocity in PSO represents the probability of the corresponding dimension taking value "1" [11]. In terms of feature selection, the velocity represents the probability of a feature being selected. Therefore, the m features with the highest velocity in a certain cluster should have the largest probability to be selected.

3.3 An Example

Taking a cluster with 30 features as an example, the following steps show the process of the proposed Gaussian position updating procedure. The elements in the position that correspond to other clusters are updated in the same way.

- Step 1: Build the Gaussian function $g(m)$ using $\mu = 1$ and $\sigma = log(10 \times 30)$;
- Step 2: Build a set of desired features DF: add feature i to the desired feature list if a random value is smaller than $\frac{1}{1+e^{-v_{id}}}$;
- Step 3: Calculate the $g(m)$ values with $m = 1, 2, 3, ...|DF|$ and normalise them;
- Step 4: Based on the normalised $g(m)$ values, the "roulette wheel selection" is performed to determine the value of m;
- Step 5: Update the position of the m dimensions with the largest velocities to "1" and all other dimensions in the same cluster to "0".

Based on the proposed Gaussian updating mechanism, we develop a new PSO approach (named GPSO) to incorporate the statistical clustering information to address feature selection problems.

4 Experimental Design

A set of experiments have been conducted to examine the performance of the proposed algorithm (GPSO). Eight benchmark datasets shown in Table 1 were chosen from the UCI machine learning repository [26], which have different numbers of features, classes and instances. The instances in each dataset are split randomly into a training set (70%) and a test set (30%). The statistical clustering method used here was taken from a recently developed algorithm [9, 10], which is not described here due to the page limit.

Table 1. Datasets

Dataset	No. of features	No. of clusters	No. of classes	No. of instances
Wine	13	6	3	178
Vehicle	18	6	4	846
Ionosphere	34	11	2	351
Sonar	60	12	2	208
Musk1	166	14	2	476
Arrhythmia	279	15	16	452
Madelon	500	11	2	4400
Multiple Features	649	15	10	2000

A small number (less than 500) of training instances are used in the statistical clustering method to speed up the clustering process, which is part of the training set on datasets such as Madelon. The number of clusters obtained are listed in the third column of Table 1.

A standard BPSO based feature selection algorithm (PSOFS), which does not consider the statistical clustering information as GPSO, is used as a baseline algorithm to test the performance of GPSO. In all the two PSO based methods, K-Nearest Neighbour (KNN) with K=5 is used in the fitness function to evaluate the classification accuracy of the selected features. The parameters are set as follows [3]: $w = 0.7298$, $c_1 = c_2 = 1.49618$, $v_{max} = 6.0$, the population size is 30, the maximum number of iterations is 100 and the fully connected topology is used. The algorithms have been conducted for 40 independent runs on each dataset. The non-parametric statistical significance test, Wilcoxon test, is performed between the testing classification performance of a PSO algorithm and all features. The significance level is selected as 0.05 (or confidence interval is 95%).

To further examine the performance of the proposed algorithms, we also compare them with two traditional feature selection methods, which are linear forward selection (LFS) [27] and greedy stepwise backward selection (GSBS). LFS and GSBS were derived from two typical feature selection algorithms, i.e. sequential forward selection (SFS) and sequential backward selection (SBS), respectively. LFS [27] restricts the number of features that are considered in each step of the forward selection, which can reduce the number of evaluations. Therefore, LFS is computationally less expensive than SFS and can obtain good results. The greedy stepwise feature selection algorithm implemented in Weka [28] can move either forward or backward. Given that LFS performs a forward selection, a backward search is chosen in greedy stepwise search to form a greedy stepwise backward selection (GSBS). GSBS starts with all available features and stops when the deletion of any remaining feature reduces the classification accuracy.

5   Results and Discussions

Table 2 shows the experimental results of PSOFS, GPSO, where "All" means that all of the available features are used for classification, "AveSize" shows the average number of features selected in the 40 independent runs, "AveAcc", "BestAcc" and "StdAcc" shows the average, the best and the standard deviation of the 40 testing accuracies. "Test" shows the results of the Wilson significance tests, where "+" (-) means PSOFS,

Table 2. Experimental Results

Dataset	Method	AveSize	BestAcc	AveAcc ± StdAcc	Test	Time
Wine	All	13	76.54			
	PSOFS	8.32	97.53	95.96 ± 1.8725	+	0.25
	GPSO	5.38	98.77	96.7 ± 2.7521	+	0.18
Vehicle	All	18	83.86			
	PSOFS	9.28	85.83	84.3 ± 0.6194	+	8.13
	GPSO	8.92	85.24	84.26 ± 0.5962	+	4.51
Ionosphere	All	34	83.81			
	PSOFS	10.38	93.33	89.05 ± 1.8444	+	1.36
	GPSO	7.5	94.29	89.26 ± 1.6631	+	0.92
Sonar	All	60	76.19			
	PSOFS	24.72	87.3	79.52 ± 2.9222	+	0.75
	GPSO	17.75	87.3	78.49 ± 3.7217	+	0.68
Musk1	All	166	83.92			
	PSOFS	83.6	89.51	85.65 ± 2.102	+	10.09
	GPSO	39.6	89.51	84.91 ± 2.5641	=	3.56
Arrhythmia	All	279	94.46			
	PSOFS	119.35	95.14	94.57 ± 0.3351	=	11.82
	GPSO	45.9	95.7	94.86 ± 0.355	+	3.83
Madelon	All	500	70.9			
	PSOFS	244.68	78.85	76.83 ± 1.2334	+	866.47
	GPSO	36.25	87.82	85.61 ± 1.0066	+	137.67
Multiple Features	All	649	98.63			
	PSOFS	295.52	99.2	99 ± 0.0962	+	726.19
	GPSO	92.25	99.27	99.02 ± 0.1258	+	112.94

GPSO is significantly better (or worse) than "All", and "=" means they are similar (no significant difference). The last column shows the average computational time used by the two PSO algorithms in a single run, which is expressed in minutes.

5.1   Results of GPSO

According to Table 2, it can be seen that the feature subsets selected by GPSO achieved significantly higher classification accuracy than using all features on **all** datasets. Furthermore, on **all** datasets, GPSO selected fewer than half of the original features, which is less than 20% on the datasets with a large number of features, i.e. the Arrhythmia, Madelon and Multiple Features datasets. For example, on the Madelon dataset, GPSO selected on average only around 7.2% of the original features (36.08 out of 500) and increased the classification accuracy from 70.9% to on average 85.61%.

Compared with PSOFS which does not use the statistical clustering information, it can be seen that GPSO achieved similar or even better classification performance than PSOFS, but the average number of features selected by GPSO is smaller or much smaller than PSOFS in **all** datasets. On the three datasets with more than 200 features, i.e. Arrhythmia, Madelon and Multiple Features, GPSO further reduced more than 60% of the feature selected by PSOFS, but still achieved slightly better classification performance than PSOFS. The reason is that on such large datasets, GPSO further removed redundant and irrelevant features, which reduced the complexity of the problem and increased the classification performance on unseen test data.

The results suggest that by developing the new Gaussian based updating mechanism in PSO, GPSO can successfully use the statistical clustering information to address feature selection problems. GPSO reduced the dimensionality of the datasets and

Table 3. Further Comparisons

Method	Wine		Vehicle		Ionosphere		Sonar	
	Size	Accuracy	Size	Accuracy	Size	Accuracy	Size	Accuracy
LFS	7	74.07	9	83.07	4	86.67	3	77.78
GSBS	8	85.19	16	75.79	30	78.1	48	68.25

Method	Musk1		Arrhythmia		Madelon		Multiple Features	
	Size	Accuracy	Size	Accuracy	Size	Accuracy	Size	Accuracy
LFS	10	85.31	11	94.46	7	64.62	18	99.0
GSBS	122	76.22	130	93.55	489	51.28		

simultaneously increased the classification performance in **all** cases, and also outperformed the standard PSO based feature selection algorithm, PSOFS.

5.2 Comparisons on Computational Time

According to Table 2, it can be seen that GPSO finished the evolutionary training process within 6 minutes in almost all cases, except on the Madelon and Multiple Features datasets, where a large number of features and instances are involved. Since the number of features selected by GPSO is much smaller than all the original features, the testing classification time will also be significantly reduced over using all the original features.

GPSO used a much shorter time than PSOFS on **all** datasets. The main reason is that as wrapper approaches, their computational time was mainly spent on evaluating the classification performance of the selected features, where a small number of features used a shorter time than a large number of features. GPSO selected a much smaller number of features than PSOFS, so its evaluations are much faster than PSOFS, especially on the large datasets. **Note** that although GPSO involves the statistical clustering process, this process is very fast since it is only performed on a part of the training examples. The computational time used by PSOFS is longer than the total time used by the statistical clustering method and GPSO.

5.3 Further Comparisons with Traditional Methods

Both LFS and GSBS are deterministic algorithms and only a single solution is obtained on each dataset, where the results are shown in Table 3. The results of using GSBS on the Multiple Features dataset are not available because the dataset is too big and the training process took too long time to finish.

Comparing Table 3 with Table 2, it can be seen that the number of features selected by LFS is usually smaller than GPSO, but GPSO achieved significantly better classification performance than LFS on almost all datasets. GPSO outperformed GSBS in terms of both the number of features and the classification performance on **all** datasets.

The results show that GPSO based on PSO and the feature clustering information can better explore the solution space to obtain better feature subsets than LFS and GSBS. In terms of the computational time, GPSO is slower than LFS because LFS selected a smaller number of features, but it is faster than GSBS on datasets with a relative large number of features.

6   Conclusions and Future Work

The goal of this paper was to develop a new approach to using the statistical clustering information in PSO for feature selection. The goal was successfully achieved by developing a new Gaussian based updating mechanism to propose a new algorithm named GPSO. GPSO was examined and compared with two traditional feature selection algorithms (LFS and GSBS) and a standard PSO based feature selection algorithm (PSOFS) on eight benchmark datasets of varying difficulty. The results show that GPSO can successfully use the statistical clustering information to select a small subset of features and achieve similar or significantly better classification performance than using all features on **all** the eight datasets. GPSO achieved significantly better classification performance than LFS, although the number of features is slightly larger. It outperformed GSBS in terms of both the number of features and classification accuracy. GPSO achieved similar classification performance to PSOFS, but selected a much smaller number of features and used a much shorter time. Compared with the original features, GPSO achieved significantly better classification performance, and reduced the number of features to an order of magnitude on the large datasets.

This work shows that statistical clustering information can be successfully used to improve the performance of a PSO based feature selection algorithm. The successes of GPSO provides motivations to further explore the use of statistical methods with evolutionary computation techniques to solve feature selection problems. For example, statistical clustering information and PSO can be used for multi-objective feature selection or for feature construction.

References

1. Dash, M., Liu, H.: Feature selection for classification. Intelligent Data Analysis 1(4), 131–156 (1997)
2. Kennedy, J., Eberhart, R.: Particle swarm optimization. In: IEEE International Conference on Neural Networks, vol. 4, pp. 1942–1948 (1995)
3. Shi, Y., Eberhart, R.: A modified particle swarm optimizer. In: IEEE International Conference on Evolutionary Computation (CEC 1998), pp. 69–73 (1998)
4. Engelbrecht, A.P.: Computational intelligence: An introduction, 2nd edn. Wiley (2007)
5. Chuang, L.Y., Chang, H.W.: Improved binary PSO for feature selection using gene expression data. Computational Biology and Chemistry 32(29), 29–38 (2008)
6. Lane, M., Xue, B., Liu, I., Zhang, M.: Particle swarm optimisation and statistical clustering for feature selection. In: Cranefield, S., Nayak, A. (eds.) AI 2013. LNCS, vol. 8272, pp. 214–220. Springer, Heidelberg (2013)
7. Cervante, L., Xue, B., Shang, L., Zhang, M.: A multi-objective feature selection approach based on binary pso and rough set theory. In: Middendorf, M., Blum, C. (eds.) EvoCOP 2013. LNCS, vol. 7832, pp. 25–36. Springer, Heidelberg (2013)
8. Bach, F.R., Jordan, M.I.: A probabilistic interpretation of canonical correlation analysis. Technical report (2005)
9. Pledger, S., Arnold, R.: Multivariate methods using mixtures: correspondence analysis, scaling and pattern detection. Computational Statistics and Data Analysis (2013), http://dx.doi.org/10.1016/j.csda.2013.05.013
10. Matechou, E., Liu, I., Pledger, S., Arnold, R.: Biclustering models for ordinal data. Presentation at the NZ Statistical Assn. Annual Conference, University of Auckland (2011)

11. Kennedy, J., Eberhart, R.: A discrete binary version of the particle swarm algorithm. In: IEEE International Conference on Systems, Man, and Cybernetics, Computational Cybernetics and Simulation, vol. 5, pp. 4104–4108 (1997)
12. Xue, B., Zhang, M., Browne, W.: Particle swarm optimization for feature selection in classification: A multi-objective approach. IEEE Transactions on Cybernetics 43(6), 1656–1671 (2013)
13. Zhu, Z.X., Ong, Y.S., Dash, M.: Wrapper-filter feature selection algorithm using a memetic framework. IEEE Transactions on Systems, Man, and Cybernetics, Part B: Cybernetics 37(1), 70–76 (2007)
14. Neshatian, K., Zhang, M., Andreae, P.: Genetic programming for feature ranking in classification problems. In: Li, X., Kirley, M., Zhang, M., Green, D., Ciesielski, V., Abbass, H.A., Michalewicz, Z., Hendtlass, T., Deb, K., Tan, K.C., Branke, J., Shi, Y. (eds.) SEAL 2008. LNCS, vol. 5361, pp. 544–554. Springer, Heidelberg (2008)
15. Kanan, H.R., Faez, K.: An improved feature selection method based on ant colony optimization evaluated on face recognition system. Applied Mathematics and Computation 205(2), 716–725 (2008)
16. He, X., Zhang, Q., Sun, N., Dong, Y.: Feature selection with discrete binary differential evolution. In: International Conference on Artificial Intelligence and Computational Intelligence (AICI 2009), vol. 4, pp. 327–330 (2009)
17. Al-Ani, A., Alsukker, A., Khushaba, R.N.: Feature subset selection using differential evolution and a wheel based search strategy. Swarm and Evolutionary Computation 9, 15–26 (2013)
18. Xue, B., Zhang, M., Browne, W.: Novel initialisation and updating mechanisms in pso for feature selection in classification. In: Esparcia-Alcázar, A.I. (ed.) EvoApplications 2013. LNCS, vol. 7835, pp. 428–438. Springer, Heidelberg (2013)
19. Wang, X., Yang, J., Teng, X., Xia, W.: Feature selection based on rough sets and particle swarm optimization. Pattern Recognition Letters 28(4), 459–471 (2007)
20. Fdhila, R., Hamdani, T., Alimi, A.: Distributed mopso with a new population subdivision technique for the feature selection. In: International Symposium on Computational Intelligence and Intelligent Informatics, pp. 81–86 (2011)
21. Yang, C.S., Chuang, L.Y., Li, J.C.: Chaotic maps in binary particle swarm optimization for feature selection. In: IEEE Conference on Soft Computing in Industrial Applications (SMCIA 2008), pp. 107–112 (2008)
22. Xue, B., Zhang, M., Browne, W.N.: Multi-objective particle swarm optimisation (pso) for feature selection. In: Genetic and Evolutionary Computation Conference (GECCO 2012), Philadelphia, PA, USA, pp. 81–88. ACM (2012)
23. Javani, M., Faez, K., Aghlmandi, D.: Clustering and feature selection via pso algorithm. In: International Symposium on Artificial Intelligence and Signal Processing, pp. 71–76 (2011)
24. Jakub Segen, J.: Feature selection and constructive inference. In: Proceedings of Seventh International Conference on Pattern Recognition, pp. 1344–1346 (1984)
25. Kira, K., Rendell, L.A.: A practical approach to feature selection. In: Assorted Conferences and Workshops, pp. 249–256 (1992)
26. Bache, K., Lichman, M.: UCI Machine Learning Repository (2013)
27. Gutlein, M., Frank, E., Hall, M., Karwath, A.: Large-scale attribute selection using wrappers. In: IEEE Symposium on Computational Intelligence and Data Mining (CIDM 2009), pp. 332–339 (2009)
28. Witten, I.H., Frank, E.: Data Mining: Practical Machine Learning Tools and Techniques, 2nd edn. Morgan Kaufmann (2005)

Global Optimization
of Multimodal Deceptive Functions

David Iclănzan

HEC Lausanne, Quartier UNIL-Dorigny,
Bâtiments Internef, 1015 Lausanne, Switzerland
david.iclanzan@gmail.com

Abstract. Local search algorithms operating in high-dimensional and multimodal search spaces often suffer from getting trapped in a local optima, therefore requiring many restarts. Even with multiple restarts, their search efficiency critically depends on the choice of the neighborhood structure. In this paper we propose an approach in which the need for the restarts is exploited to improve the neighborhood definitions. Namely, a graph clustering based linkage detection method is used to mine the information from several runs, in order to extract variable dependencies and update the neighborhood structure, variation operators accordingly. We show that the adaptive neighborhood structure approach enables the efficient solving of challenging global optimization problems that are both deceptive and multimodal.

1 Introduction

Generic stochastic optimization methods like evolutionary algorithms, swarm algorithms, stochastic hill climbing, simulated and quantum annealing, tabu search etc. are often used for locating in reasonable time a good approximation of the global optimum of hard problems, defined over large search spaces and/or exhibiting multimodality, deceptiveness.

However, these methods are not always applicable as there are problems that are intractable using fixed, problem independent operators and representations [1]. Efficient solving of such problems requires methods that can learn the structure of a problem on the fly and use operators that adapt the linkage between variables. In the case of population based methods the sampling offered by the elements of the population opens the way for inferring valuable statistical information. For these methods, the replacement of operators like crossover and mutation with a repeated selection – model-building – sampling process, enables the required adaptability of the neighborhood structures and adequate performance.

The family of these methods, often called competent evolutionary algorithms [2] include such examples as the fast messy genetic algorithm [3], the linkage learning genetic algorithm [4], the extended compact genetic algorithm [5], the Bayesian Optimization Algorithm (BOA) [6], SEAM [7], DevRep [8], Compact Genetic Codes [9], Hierarchical Genetic Algorithm [10], hBOA [11] and DS-GMA++ [12].

C. Blum and G. Ochoa (Eds.): EvoCOP 2014, LNCS 8600, pp. 145–156, 2014.

Incorporating statistical models in trajectory based methods like stochastic hill climbing, simulated and quantum annealing, tabu search etc. is not straightforward as these methods sample just one point of the search space at each step. However, being more lightweight and prone to premature convergence, these methods are used in conjunction with a a random restart mechanism that facilitates a better balance between exploration and exploitation of the search space. Therefore, we apply learning to the results obtained from several runs, extracting linkages and providing properly adapted search operators for future runs.

In this paper we extend the simulated annealing trajectory based stochastic optimizer with a graph clustering based linkage detection method. We investigate the proposed method performance on the concatenated trap function, a problem designed to test a global optimization method's ability to handle deceptiveness and multiomdality.

The next Section presents and discuss some preliminary notions. The proposed method is detailed in Section 3. Section 4 describes the experimental setup, followed by a discussion of the results in Section 5. Finally, the paper concludes in Section 6.

2   Preliminaries

2.1   Simulated Annealing

Simulated annealing is a generic stochastic global optimization metaheuristic [13], inspired from the annealing process, which involves heating and controlled cooling of solids in order to increase the size of their crystals and reduce their defects due to the changes in internal structure .

The simulated annealing algorithm makes an analogy between the thermodynamic free energy and the objective function. In the annealing process, a slower rate of cooling produces a bigger decrease in energy. The simulated slow, controlled cooling, effects the probability that the simulated annealing accepts a worse solution, decreasing it over time.

In the beginning, when the artificial temperature is high the algorithm will often accept solutions that are worse than the current solution, encouraging exploration, facilitating the escape from the basin of attraction of local optima. As the temperature is reduced the probability of accepting worse solutions decreases, focusing the search more on exploitation. The balance provided by the controlled, gradually decreasing artificial cooling enables the simulated annealing algorithm to perform well on large search spaces. The pseudo-code of the method is presented in Alg. 1.

The simulated annealing as many other generic metaheuristics uses a fixed neighborhood structure which can severely hinder its performance in problems that require the discovery and efficient mixing of partial sub-solutions, building-blocks. At each temperature the algorithm performs a number of iterations that is controlled by a positive parameter *nrtries*. The algorithms terminates when it finds a good enough solution, or when there are no improvements over the last *limitNI* iterations.

Algorithm 1. Simulated annealing

```
1  Define an initially high temperature T;
2  Define a cooling schedule cooling(T);
3  Define an energy function E(state);
   /* Start from a random point                                    */
4  x ← RandomState();
5  while not converged and noimprovementcount < limitNI do
6      for i ← 1 : nrtries do
7          x_new ← neighboor(x);
8          Δ ← E(x_new) − E(x);
9          if Δ < 0 then
10             x ← x_new;
11         else
               /* Evaluate if the worse state is to be accepted     */
12             if rand < e^(−Δ/T) then
13                 x ← x_new;

       /* Decrease the temperature according to the cooling schedule  */
14     T ← cooling(T);
```

2.2 Graph Clustering Based Model Building

Dependent variables are detected by applying a maximum flow clustering algorithms to an adjacency matrix which contains the pairwise interactions between variables. Therefore, the method is model evaluation free, fast, scalable and easily parallelizable.

Maximum flow clustering algorithms simulate a special flow within a graph, which promotes flow where the current is strong whilst reducing flow where the current is weak. These procedure reveal the cluster structure within the graph, as the flow across borders between different groups diminish with time, while it increases within the group.

Simulation of flow through a graph is easily done by transforming the adjacency matrix into a column-stochastic square matrix, where each column sums to 1. This matrix, denoted by M, can be interpreted as the matrix of the transition probabilities of a random walk (or a Markov chain) defined on the graph, where $M(j, i)$ represents the probability (stochastic flow) of a transition from vertex v_i to v_j. Flow expansion can be simulated by computing powers of the flow (Markov) matrix M.

One of the most well known and used maximum flow based graph clustering algorithm is the Markov Clustering Algorithm (MCL) [14]. It offers several advantages, like a simple, elegant mathematical formulation, robustness to topological noise [15], support for easy paralellization and adaptation via a simple parameter enables the obtaining of clusters of different granularity.

MCL iteratively simulates random walks within a graph by applying two operators called expansion and inflation, until convergence occurs. At the end of each inflation step a pruning step is also performed, in order to reduce the computational complexity by making M sparse. The MCL process may be regarded as alternative expansion and contraction of the flow in the graph. The expansion step is responsible in spreading the flow out of a vertex to potentially new vertices and with the strengthening of the flow to those vertices which are reachable by multiple paths. This has the effect of enhancing within-cluster flows as there are more paths between the nodes belonging to the same cluster. The inflation operator is responsible for both strengthening intra-cluster flow and weakening inter-cluster flow of current and by this, introducing a non-linearity in the distribution of the flows. At the beginning the flow distribution is relatively smooth and uniform, but with each iteration it becomes more and more peaked. In the end, all the nodes within a tightly-linked group of nodes will start to flow towards one node within the group, forming star sub-graphs associated with the MCL limits.

The idealized Markov Cluster process, consisting just from the expansion and inflation operators is known to converge quadratically in the neighborhood of so called doubly idempotent matrices [14]. In practice, the numbers of epochs until convergence is reported to be nearly always far below 100.

While several types of models can be extracted from the MCL iterants [16], here we choose a simple approach, which extracts the linkage groups from the clusters obtained after convergence of the MCL. Here, the nodes have found one "attractor" node to which all of their flow is directed, corresponding to only one non-zero entry per column in M. Nodes sharing the same "attractor" node are grouped in clusters and their corresponding variables form a linkage group that can be used by the modified simulated annealing algorithm.

3   Extended Simulated Annealing

The extended simulated annealing has two phases that are alternatively applied. In the first phase, several runs of simulated annealing are performed according to the current linkage model and the obtained result are collected in a memory. In the second phase this accumulated search experience is exploited in order to learn which variables are linked and the model is updated accordingly. Then the optimization procedure continues again with the first phase but now using a model that is more suited to the particular optimization problem. The details of these two phases are outlined in Algorithm 2.

The parameters of the algorithm are n - number of variables, n_S - number of simulated annealing runs in the search experience accumulation phase, @stopping_cond - stopping criteria which usually places a bound on the quality of the found solution and/or time, number of objective function evaluations.

$R \leftarrow corrcoef(memory)$ returns a matrix R of correlation coefficients calculated from an input matrix *memory* whose columns are variables and whose rows are the bits of the solution found by the particular simulated annealing runs.

Algorithm 2. Extended simulated annealing

    Input: n, n_S
    /* Initially each variable is independent   */
1   $M \leftarrow 1 : n$;
2   **while** *not* @*stopping_cond* **do**
        /* PHASE I - ACCUMULATE SEARCH EXPERIENCE   */
3   **for** $i = 1, n_S$ **do**
            /* Apply the simulated annealing algorithm   */
4   Define an initially high temperature T;
5   Define a cooling schedule $cooling(T)$;
6   Define an energy function $E(state)$;
            /* Start from a random point   */
7           $x \leftarrow RandomState()$;
8   **while** *not* *converged* **and** *noimprovementcount* $<$ *limitNI* **do**
              /* Perturbation according to the model M   */
9               $x\_new \leftarrow neighboor(x, M)$;
10             $\Delta \leftarrow E(x\_new) - E(x)$;
11   **if** $\Delta < 0$ **then**
12                   $x \leftarrow x\_new$;
13   **else**
                 /* Evaluate if the worse state is to be accepted   */
14   **if** $rand < e^{-\frac{\Delta}{T}}$ **then**
15                      $x \leftarrow x\_new$;
              /* Decrease the temperature   */
16             $T \leftarrow cooling(T)$;
17           $memory[i] \leftarrow x$;
        /* PHASE II - UPDATE THE MODEL   */
        /* Calculate correlation coefficients between variables   */
18       $R \leftarrow corrcoef(memory)$;
        /* Apply MCL to extract groups of dependent variables   */
19       $M \leftarrow MCL(R)$;

The matrix $R \leftarrow corrcoef(memory)$ is obtained from the covariance matrix $C \leftarrow cov(memory)$ according to the formula

$$R(i,j) = \frac{C(i,j)}{\sqrt{C(i,i)C(j,j)}} \tag{1}$$

The essence and capabilities of the proposed method lies in that the perturbation for obtaining x_new operates over a linkage model that is continuously adapted. Variables that are found to be connected are always altered together.

In the beginning each variable forms a separate linkage group, for an 8 variable problem, the initial model would be $M = \{[x_1], [x_2], \ldots [x_8]\}$. With this model the method is equivalent to the standard simulated annealing, when generating x_new each variable is perturbed according to a mutation probability p_mut.

However, if from several runs it can be determined for example that the first four, respectively the last two variables are correlated, the model is updated to reflect this knowledge: $M = \{[x_1, x_2, x_3, x_4], [x_5], [x_6], [x_7, x_8]\}$. In the subsequent runs the simulate annealing will perturb states respecting these dependencies, always altering the clustered variables together, thus eliminating sub-solution disruption and facilitating the mixing of these building-blocks.

4   Experiments

Deceptive functions are among the most challenging problems as they exhibit one or more deceitful optima located far away from the global optimum. The basins of attraction of the local-optima are much bigger than the attraction area of the optimal solutions, thus following the objective function gradient is misleading on average.

Order k deceptive trap function or simply k-Trap, is a function of unitation (its value depends only on the numbers of 1's in the input string), based on two parameters f_{high} and f_{low} which define the degree of deception and the fitness signal-to-noise ratio. The input that contains all but ones is maximally rewarded; for the other cases, the fitness of the block is directly proportional to the number of zeros, the string of all zeros being a strong local optima.

Let u be the unitation of the binary input string. The k-trap function is defined as:

$$trap_k(u) = \begin{cases} f_{high} & \text{, if } u = k; \\ f_{low} - \frac{u+1}{k} & \text{, otherwise.} \end{cases} \qquad (2)$$

Concatenating m copies of this trap function [17] gives a global additively separable, boundedly deceptive function over binary strings:

$$CTF(x) = -\sum_{i=0}^{m-1} trap_k \left(\sum_{j=ki}^{ki+k-1} x_j \right) \qquad (3)$$

In this paper we operate over 120 bit length CTF problems, concatenating $m = 20$ k-Trap functions based on $k = 6$, $f_{high} = f_{low} = 1$. For this setup, the test function have 240 local optima. The global optima, the string of all ones, has a value of -20.

In a first experiment, we tested how the simple simulated annealing performs on this 120 bit CTF function. The performance of a stochastic search method can severely depend on its parametrizations. As there are no definite guidelines on how to achieve the best configuration, we decided to test a comprehensive range of parametrizations over 20000 runs. For each run we randomly determined values for the perturbation probability ($p\_mut \in [0.01, 0.25]$), the number of maximum allowed consecutive rejections ($1000 \le limitNI \le 20000$) and the number of tries per temperature ($100 \le nrtries \le 2000$).

In the second experiment we tested the performance of the extended method averaged over 100 runs, using a perturbation probability p_mut of 0.05, the allowed maximum number of rejections $limitNI$ was set to 2000 and we performed

Parametrization of the SA runs with performance >= -15.00 (8.46% of runs)

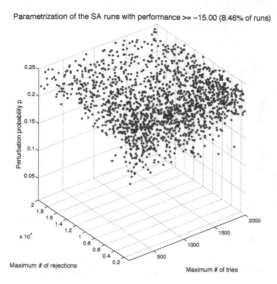

Fig. 1. Parametrizations leading to worst performance of the classical simulated annealing

just on iteration per temperature. In the search experience accumulation phase, the simulated annealing is run $n_S = 100$ times.

For both methods the cooling schedule was set to $T_{k+1} = 0.9T_k$.

5   Results

5.1   Performance of the Classical Simulated Annealing

The simple simulated annealing spent 567980353 functions evaluations in the 20000 runs. Despite the high number of trials, it was never able to find the global optimum, always being mislead by the strong deceptiveness and multimodality of the CTF.

Figure 1 depicts the parameter configurations which resulted in very poor performance for the classical simulated annealing, with objective function values greater than -15. As seen, this corresponds to parametrizations where the mutation probability is very high or both the number of allowed rejections and trials per temperature is low.

The parametrizations leading to the best performances for the classical method are shown in Figure 2. The plots in the parameter space reveal opposite trends, namely that the method requires high values for the number of allowed rejections and trials per temperature. Surprisingly, the values of the perturbation rates cover a wide spectrum, varying between 0.02 and 0.15. Represented by a dot, this figure also shows the best solution found, which had a value of -18.5. This value corresponds to solutions for which the method found the optimal

Parametrization of the SA runs with performance <= −18.00 (0.24% of runs)

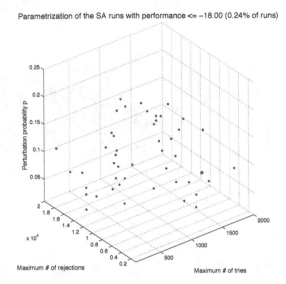

Fig. 2. Parametrizations leading to best performance of the classical simulated annealing, where the method managed to correctly identify 8-11 trap functions out of the 20. The dot denotes the best solution found, corresponding to the -18.5 fitness value.

setting of all ones for 11 k-trap sub-functions out of 20 (discovering 55% of the correct subsolutions) while converging to the deceptive optima of all zeros on the remaining 9 sub-functions.

Figure 3 depicts the histogram of the results obtained from the 20000 runs. With moderate settings for the perturbation probability the classical simulated annealing surpasses the solution of all zeros (which has a value of -16.6666) but is rarely able to resolve - find the optimal setting of all ones for half or more of the $m = 20$ k-trap sub-functions (values of -18.3333 or lower). These empirical results confirm that classical simulated annealing is not a suitable solver for the the CTF as this function requires the efficient mixing of building-blocks with deceptive intra-block gradient.

5.2   Performance of the Extended Simulated Annealing

In contrast, in our second experiment the proposed extended simulated annealing found the global optima in every single run. Figure 4 depicts the juxtaposition of the evolution of these 100 runs, their line style signaling how many learning cycles, model updates they required until convergence to the global optima. The solid lines, accounting for 95% of the cases, corresponds to runs where global optima was found after just one linkage learning phase. The remaining 5% of cases, depicted with dashed lines, corresponds to the cases where 2 or 3 neigborhood structure updates were required before convergence.

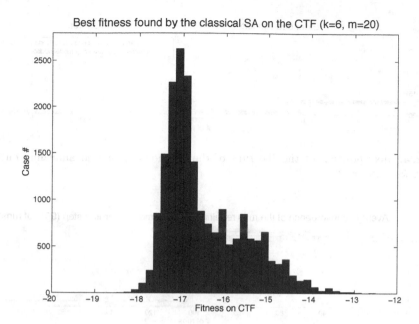

Fig. 3. Histogram of the results obtained from 20000 runs of the classical simulated annealing on CTF with $k = 6$, $m = 20$

In the first 100 epochs, when the model is the trivial one with every variable independently perturbed, the search is equivalent to the classical simulated annealing. Therefore, as expected, the performance is mediocre, the search never reaching the -18 fitness threshold.

After the first 100 runs, when the first round a linkage learning is applied we can see a huge qualitative change in performance. 95% of the time, through epochs 101-200 the search repeatedly founds the global optimum, which has the fitness value -20. This sharp performance gain is due to the adapted neighborhood structure, linkage based perturbation which maximizes the chances of solving k-trap sub-functions while also preventing the disruption of already converged sub-solutions.

However, Figure 4 also reveals that in a few instances, on the same epoch range severely degenerates. This happens in cases when the learnt model is not accurate, containing many spurious, false linkages. Following through epochs 200-400 we can observe that even in this cases, the method recovers each time, eventually learning an accurate model and converging to the global optima. Algorithms using probabilistic models had been long prone to the issue of premature convergence, where the methods can not recover from an inadequate model derived in the early phases from a biased sample due insufficient sampling. As the observed runs suggest, the continuous cycling and transition between exploration and exploitation provided by the simulated annealing search strategy seem to alleviate this problem.

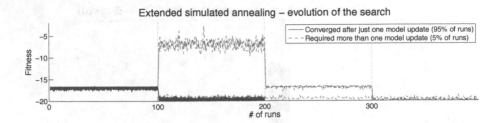

Fig. 4. Juxtaposition of the 100 runs of the extended simulated annealing on CTF with $k = 6$, $m = 20$

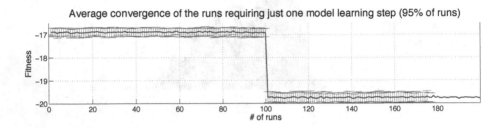

Fig. 5. Evolution of the average performance of the extended simulated annealing on CTF with $k = 6$, $m = 20$. After epoch 100, following the linkage adaptation, we observe a huge qualitative increase in the average performance, while standard deviation remains the same.

For the runs requiring just one linkage learning round, the average convergence rate along with its standard deviation is depicted in Figure 5. Again, the huge gain in performance can be observed immediately after epoch 100, when the neighborhood structure is adapted according to the linkage groups revealed by the graph clustering algorithm. Aside from the greatly shifted average performance, the extended simulated annealing exhibits the same search dynamics (stable, almost constant average, same standard deviation) before and after the neighborhood adaptation.

Figure 6 depicts the required number of objective function evaluations until convergence of the proposed method. The cases above the box are the outliers, where the method required more than one model update due to the spurious linkages incorporated in the first model. The average on the figure is located at 665702.35 evaluations, with a standard deviation of 229118.249553. The quickest convergence took 579841 evaluations while the longest one required 1872947.

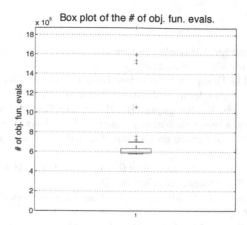

Fig. 6. Box-plot of the required number of CTF evaluations by the extended simulated annealing

6   Conclusions

Problems requiring the discovery and mixing of sub-solutions often can not be solved by problem independent search operators that while optimizing one sub-problem may disrupt already converged sub-solutions.

In this paper we extended the simulated annealing stochastic optimizer with a neighborhood structure adapting mechanism, linkage based perturbation. The linkages, groups of dependent variables are extracted from a correlation coefficient matrix, describing the pairwise dependencies between variables by means of MCL, a maximum flow based graph clustering algorithm. The proposed method showed a very robust behavior, 100% success rate on a problem used to test a method ability to overcome deceptiveness and multimodality. Comprehensive test runs showed that the same problem can not be solved by a simulated annealing using fixed neighborhood structure.

Also, the experiments revealed a very promising phenomena, namely that the extended simulated annealing, by alternating exploration and exploitation phases can recover from erroneous linkage models. This opens the way to study problems where other model based search methods struggle with the issue of premature convergence as they can not reco.

Future work will also experiment with the incorporation of richer, probabilistic models and will address other classes of optimization problems.

Acknowledgments. We acknowledge the support of the Sapientia Institute for Research Programs (KPI).

References

1. Pelikan, M.: Hierarchical Bayesian optimization algorithm: Toward a new generation of evolutionary algorithms. Springer (2005)
2. Goldberg, D.E.: The Design of Innovation: Lessons from and for Competent Genetic Algorithms. Kluwer Academic Publishers, Norwell (2002)
3. Goldberg, D.E., Deb, K., Kargupta, H., Harik, G.: Rapid, accurate optimization of difficult problems using fast messy genetic algorithms. In: Proceedings of the Fifth International Conference on Genetic Algorithms, San Mateo, CA, pp. 56–64. Morgan Kaufman (1993)
4. Harik, G.R., Goldberg, D.E.: Learning linkage. In: Belew, R.K., Vose, M.D. (eds.) FOGA, pp. 247–262. Morgan Kaufmann (1996)
5. Harik, G.R., Lobo, F.G., Goldberg, D.E.: The compact genetic algorithm. IEEE-EC 3(4), 287 (1999)
6. Pelikan, M., Goldberg, D.E., Cantú-Paz, E.: BOA: The Bayesian optimization algorithm. In: Banzhaf, W., et al. (eds.) GECCO 1999, Orlando, FL, July 13-17, vol. I, pp. 525–532. Morgan Kaufmann Publishers, San Fransisco (1999)
7. Watson, R.A., Pollack, J.: A computational model of symbiotic composition in evolutionary transitions. Biosystems 69(2-3), 187–209 (2003), Special Issue on Evolvability, ed. Nehaniv
8. de Jong, E.D.: Representation Development from Pareto-Coevolution. In: Cantú-Paz, E., et al. (eds.) GECCO 2003. LNCS, vol. 2723, pp. 262–273. Springer, Heidelberg (2003)
9. Toussaint, M.: Compact Genetic Codes as a Search Strategy of Evolutionary Processes. In: Wright, A.H., Vose, M.D., De Jong, K.A., Schmitt, L.M. (eds.) FOGA 2005. LNCS, vol. 3469, pp. 75–94. Springer, Heidelberg (2005)
10. de Jong, E.D., Thierens, D., Watson, R.A.: Hierarchical genetic algorithms. In: Yao, X., et al. (eds.) PPSN VIII. LNCS, vol. 3242, pp. 232–241. Springer, Heidelberg (2004)
11. Pelikan, M., Goldberg, D.E.: Escaping hierarchical traps with competent genetic algorithms. In: Spector, L., et al. (eds.) GECCO 2001, July 7-11, pp. 511–518. Morgan Kaufmann, San Francisco (2001)
12. Yu, T.L., Goldberg, D.E.: Conquering hierarchical difficulty by explicit chunking: substructural chromosome compression. In: GECCO 2006, pp. 1385–1392. ACM Press, NY (2006)
13. Kirkpatrick, S., Gelatt, C.D., Vecchi, M.P.: Optimization by simulated annealing. Science 220, 671–680 (1983)
14. van Dongen, S.: Graph Clustering by Flow Simulation. PhD thesis, U. of Utrecht (2000)
15. Brohée, S., van Helden, J.: Evaluation of clustering algorithms for protein-protein interaction networks. BMC Bioinformatics 7, 488 (2006)
16. Iclănzan, D., Dumitrescu, D.: Graph clustering based model building. In: Schaefer, R., Cotta, C., Kołodziej, J., Rudolph, G. (eds.) PPSN XI. LNCS, vol. 6238, pp. 506–515. Springer, Heidelberg (2010)
17. Deb, K., Goldberg, D.E.: Analyzing deception in trap functions. In: Whitley, L.D. (ed.) Foundations of Genetic Algorithms 2, San Mateo, pp. 93–108. Morgan Kaufmann (1993)

Learning Inherent Networks
from Stochastic Search Methods

David Iclănzan, Fabio Daolio, and Marco Tomassini

Faculty of Business and Economics, University of Lausanne, Switzerland
david.iclanzan@gmail.com, {fabio.daolio,marco.tomassini}@unil.ch

Abstract. Analysis and modeling of search heuristics operating on complex problems is a difficult albeit important research area. Inherent networks, i.e. the graphs whose vertices represent local optima and the edges describe the weighted transition probabilities between them, enable a network characterization of combinatorial fitness landscapes. Methods revealing such inherent structures of the search spaces in relation to deterministic move operators, have been recently developed for small problem instances. This work proposes a more general, scalable, data-driven approach, that extracts the transition probabilities from actual runs of metaheuristics, capturing the effect and interplay of a broader spectrum of factors. Using the case of NK landscapes, we show that such an unsupervised learning approach is successful in quickly providing a coherent view of the inherent network of a problem instance.

1   Introduction

Knowledge of the fitness landscape structure of hard problem instances, also called the search space, is an important issue in approximately solving those problems since the performance of metaheuristics on the problem crucially depends on this structure [1]. Empirical evaluation of the search space properties is usually carried out by means of simple metrics such as fitness-distance correlation or autocorrelation length. However, as each of those only captures a single facet of the fitness landscape, recent efforts are targeted on developing richer, more expressive models (see e.g. [2]).

The search space is fully defined by giving the (finite) set S of admissible solutions, an objective or fitness function that measures the quality of the solutions, and a neighborhood structure that, for each $x \in S$, tells which are the solutions that can be reached from x by the application of a pre-defined move operator. In previous work [3,4], attempts have been made to provide a compact view of the complexity of the search space by defining a derived directed network whose nodes are the optima of the search space and whose edges are approximate transition probabilities between those optima and thus, implicitly, between the corresponding basins of attraction. These networks have been variously called "inherent networks" or "local optima networks" (LONs) and have their origins in Monte Carlo studies of the energy of molecular clusters and macromolecules [5].

C. Blum and G. Ochoa (Eds.): EvoCOP 2014, LNCS 8600, pp. 157–169, 2014.

In contrast to energy landscape, though, in combinatorial spaces the mapping of states to energy levels is not continuous, thus the notions of minima and transitions have been readapted. In all previous work about LONs, local optima are defined according to one-move hill-climbers. Transitions, then, can be defined by exhaustively sampling all the solutions in a basin of attraction and looking at the basin of their one-mutant neighbors (*basin-transition* edges [3,4]), or by performing a number of kick-moves from a local optimum and looking at the basin of the corresponding new solutions (*escape* edges [6]). Both models provide a compressed description of the combinatorial landscape that is relevant for local search heuristics. It has been shown that some statistical features of these complex networks correlate with problem difficulty, especially their mean path lengths, the out-degree, and the lengths of paths to the global optimum, which help to determine whether a given search heuristic will perform well on them [7].

While the relevance of inherent networks analysis has been assessed, the current LON extraction methods have some limitations we wish to address:

Scalability: the computation of the basin-transition and escape edges have exponential running times, which limit the applicability of the approach built upon them to smaller problem dimensions.

Generality: these above methods computed transitions, built the inherent network in relation with well-defined neighborhood structures resulting from simple deterministic move operators (1 or 2-bit flip). With stochastic operators, though, the notion of basin adjacency is not well-defined. Popular metaheuristics often use several stochastic move operators that induce complex neighborhood structures.

Abstraction Level: as recent results showed [8], search heuristics using the same move operator but slightly different exploitation strategies (first improvement hill-climbing versus best improvement hill-climbing), result in different inherent networks. This strongly suggests that inherent network modeling is more suited at the metaheuristics-problem instance level rather than at the move operator-problem instance one.

For these reasons, we depart towards a data-driven approach that extracts the inherent network from many runs of a stochastic search method. We use a Self Organizing Map to obtain a compressed view of the search space, which facilitates the monitoring of the approximate regions visited by the metaheuristic. By mapping and tracking the sequence in which nodes are visited, we obtain an approximation of the transition probabilities characterizing the inherent network.

The paper is organized as follows: Section 2 gives the details of the proposed method. Section 3 describes the experimental setup, while results are reported in Section 4. Section 5 concludes the paper, also outlining future lines of research.

Algorithm 1. Stochastic hill-climber

Input: N, $maxnoimprovement$, D
Output: $state$
1 $state \leftarrow RandomBinaryState(N)$;
2 $noimprovement \leftarrow 0$;
3 $epoch \leftarrow 0$;
4 **while** $noimprovement < maxnoimprovement$ **do**
5      $epoch \leftarrow epoch + 1$;
6      $newstate \leftarrow state$;
7   **for** $i=1{:}N$ **do**
8   **if** $rand < 1/N$ **then**
9   bitflip($newstate[i]$);

      /* call needed for basin adjacency computation   */
10   updateAdjacencyMatrix(state, epoch, D);
11   **if** $Energy(newstate) > Energy(state)$ **then**
12          $state \leftarrow newstate$;
13          $noimprovement \leftarrow 0$;
14   **else**
15          $noimprovement \leftarrow noimprovement + 1$;

---

2 Methods

2.1 Studied Problem and Search Heuristic

For easy comparison purposes, we use the same 18-bit NK landscape model as in previous work on this topic [3,4,6,7]. NK landscapes are a model of stochastic binary landscapes that can be tuned from smooth to rugged according to a single parameter K that defines the problem non-linearity [9]. In the model, N refers to the length of the bit strings; each bit of the N sites on a string contributes to its fitness value but the value of each contribution depends on K other positions from the bit-string. For $K = 0$ we have an "easy", unimodal landscape; for $K = N - 1$ a "hard", highly-rugged and completely uncorrelated one.

In order to be able to compare with [6], as a search heuristic we use a simple stochastic hill-climber, which at each iteration flips the bits of the actual state with a probability of $1/N$, where N is the dimensionality of the search space. The newly generated state is accepted only if it improves the current state. The search stops after a predefined number of consecutive trials without improvement. The pseudocode of the stochastic hill-climber is outlined in Algorithm 1. The role of the input matrix D and the procedure call from line 10, are particular to our proposed method and are later detailed in Section 2.4 in Procedure **updateAdjacencyMatrix**. Except for those, Algorithm 1 is the much general $(1 + 1)$ES with $1/N$ mutation that is widely used in theoretical studies.

Mapping the search dynamics of metaheuristics over combinatorial landscapes to networks, requires the following components: (i) a way to determine the set of nodes; (ii) a method that can approximate the search progress through the basins of attractions corresponding to the nodes; (iii) measurement and quantification of the transfer probabilities between basins of attraction. We detail these steps in the following.

2.2   Monitoring the Search Dynamics

Stochastic metaheuristics dynamics are hard to follow due to their ergodic nature. Usually in each step they can theoretically reach an exponential number of states, making an exact tracking infeasible.

In order to obtain a reduction in dimensionality, our proposed approach uses Self-Organizing Maps (SOM) [10] to follow and mine data from the search process of metaheuristics. SOM develop a topology-preserving, discrete, low-dimensional representation of a higher-dimensional space through unsupervised learning. SOM have been successfully applied in a wide range of research areas covering data mining, pattern recognition, analysis and control of complex systems. In the field of metaheuristics, SOM have been used to visualize the trajectory of an individual in the search space [11], and to map algorithm performance in the space of problem instances [12]. In relation to binary search spaces, SOM have been successfully applied in genetic algorithms to mine information about the search dynamics [13]. In that work, the information provided by the map was used to intelligently balance exploration and exploitation in these stochastic, population-based methods.

In this paper, the SOM enables to track the search trajectory across points of interests, the basins of local optima. After preliminary tests, a SOM of 4096 neurons arranged on a 64 by 64 grid was employed. Its training still limits the scalability of the approach, but has to be carried out only once for all the instances of the same size. We used 3000 training epochs, in each epoch presenting to the network in random order the entire 18-bit binary input space. After training, each one of the 4096 neurons maps a particular region of the 18-bit solution space, providing a compressed view.

The trained SOM is used to build a simple function that approximates if a solution belongs to the basin of a local optimum of interest. The idea is that if two solutions resemble topologically, matching to the same SOM neuron, and their energy difference is small, being under a predefined threshold, then there is a high probability that they belong to the same basin of attraction. With the help of this function we can efficiently monitor the search process in relation to basins of attraction it passes through with time, being able to record the sequence and the frequency with which basins of attractions are visited.

The pseudocode of the identification is presented in the Function named **isBasinOf**. The approximation is noisy, i.e. the basin identifying function can give false positives. To address this issue, we sample a large number of search trajectories. This helps to reduce the effect of the noise and correctly highlights the trends in which basins of attraction tend to follow each other.

Function isBasinOf(node, state, energythreshold)

Result: boolean
/* compute the best matching units (bmu) of the target node and
current state   */
1 $bmu\_node \leftarrow SOM(node)$;
2 $bmu\_state \leftarrow SOM(state)$;
/* compute the energy delta   */
3 $\Delta \leftarrow$ Energy($node$) - Energy($state$);
4 **if** $(bmu\_node = bmu\_state)$ **and** $(\Delta < energythreshold)$ **then**
5 $\quad$ | return **true**;
6 **else**
7 $\quad$ | return **false**;

2.3 The Nodes

Transition frequencies can be measured over arbitrary set of solutions. However, following the analogy with inherent networks of energy landscapes, solutions corresponding to network nodes should be strong solutions, possibly local optima relative to the search heuristic. The source of these points can be external, known good solutions provided by human experts or reported by other search methods. If this is not available, they can be determined internally by running the heuristics many times and choosing the best points where the search plateaued.

To allow for a direct comparison, in this paper we use as nodes the local optima reported and analysed in previous work [7]. For each studied problem instance, we restrict our analysis to the top local optima that map to different neurons in the trained SOM. For high K, the NK landscape has an exponential number of local optima. For these instances we cap the number of nodes to the top 1024 local optima in order to maintain a minimum ratio of 1:4 between the nodes of the inherent networks and the neurons from the trained SOM.

2.4 The Edges

With the help of the above described **isBasinOf** Function, in each step of the stochastic hill-climber we check if the newly generated state belongs to a basin of attraction of one of the nodes. If yes, for all the other basins that have been visited before, the adjacency matrix describing the empirical transfer frequencies between basins of different local optima is updated. The amount of update is proportional to the number of epochs elapsed between the visits, being 1 for consecutive visits and tending to 0 as the length of elapsed epochs is closer to the *maxnoimprovement* threshold. The update process is outlined in Procedure **updateAdjacencyMatrix**.

As seen in line 1 in Procedure **updateAdjacencyMatrix**, we use an arbitrarily value of 0.01 for the *energythreshold* variable in Function **isBasinOf**, which accounts for 2% to 10% of the fitness variance, depending on the instance.

Procedure updateAdjacencyMatrix(state, epoch, D)

---

1 $energythreshold \leftarrow 0.01$;
2 **for** $b \leftarrow 1 : nrOfNodes$ **do**
   ```
   /* check if the current search state maps to any basin b        */
   ```
3   **if** $isBasinOf(nodes[b], state, energythreshold)$ **then**
4   **for** $i \leftarrow 1 : nrOfNodes$ **do**
5              $l \leftarrow lastTimeVisted(nodes[i])$;
   ```
   /* if other basins i have been visited before, update the
   adjacency matrix                                               */
   ```
6   **if** $l > 0$ **then**
   ```
   /* the update is weighted with the elapsed epochs
   between the visits                                             */
   ```
7                  $t \leftarrow min(epoch - l, maxnoimprovement)$;
8                  $D(i,b) \leftarrow D(i,b) + \frac{maxnoimprovement - t + 1}{maxnoimprovement}$;

The data-driven approach needs many runs in order to have time to discover and reinforce (thus, diminish the impact of false positives) true basin adjacency trends. For each studied NK landscape instance, first we initialize the adjacency matrix D to all zeros, then we perform 10000 stochastic hill-climber runs as defined in Algorithm 1 to mine and accumulate the transfer frequencies into D.

In the last step we normalize D by dividing each non-zero $D(i,j)$ entry by the sum of the elements on the ith row. This results in a stochastic adjacency matrix where the $D(i,j)$ entries describe the probability that the search will move from the basin of attraction corresponding to the ith node to the basin of the jth node. Note that, from Procedure **updateAdjacencyMatrix**, such transitions need not to be direct jumps as in [6], rather they can be interpreted as conditional probabilities that the search might end up in the jth node some time after having visited the ith node, with transition weight relating to the time interval between visits.

3 Experiments

With N fixed to 18, as in previous studies, the ruggedness of the NK landscape was increased from $K = 2$ to $K = 16$ by steps of 2. For statistical significance, for each K we generated 100 different random NK landscape instances. On each landscape we ran the classical 2-bit flip analysis [6,7] for baseline comparison and the proposed method with 3 different parametrizations of the stochastic hill-climber, which had the $maxnoimprovement$ condition set to 180, 640, and finally 1440 steps. This parameter regulates how much exploration time the method has; with an increased value it is expected that the hill-climber is able to reach basins of attraction that are further away in a Hamming distance sense.

The previous and the present methods used for deriving the inherent networks significantly differ, the first being based on deterministic move operators

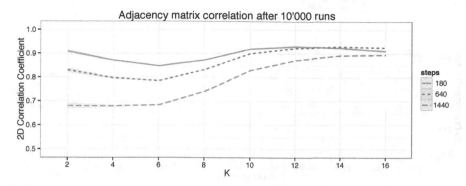

Fig. 1. Correlation between normalized adjacency matrices corresponding to inherent networks obtained by the deterministic method mentioned in Section 1 and our proposed data-driven approach. Results are averaged on 100 independent landscapes, shaded areas (where visible) give the standard error on the mean. Line type and color refer to the stopping criterion of the stochastic hill-climber (see legend on the right).

[6,7] while the second extracts the network from stochastic search trajectories. However, as both approaches tend to sample close points in a Hamming distance sense, for a first assessment we performed a direct comparison: the obtained transition matrices were normalized (non-zero entries are divided by the row sum) and their corresponding Pearson 2D matrix correlation coefficient were computed. In a second stage we analyzed how the inherent networks obtained by our method correlate with the parametrized non-linearity of the NK landscapes.

4   Results

4.1   Comparative and Convergence Analysis of the Inherent Networks

The results of the direct comparison are depicted in Figure 1. The inherent networks obtained by the data-driven approach from 10000 stochastic hill-climber runs, significantly correlate with the deterministic approach for all K. The results are very similar in all the 100 different runs for each K, resulting in very small standard errors of the mean. As seen in the figure, the correlation is weaker for smaller K, as in these problem instances the landscape is smoother and with fewer basins of attraction. Here, the stochastic hill-climber is often able to escape basins of attraction, moving towards stronger solutions. The discrepancy is directly proportional to the exploration time, value of the *maxnoimprovement* parameter, entitled to the stochastic hill-climber. For high values of K, the landscapes are highly uncorrelated, with an exponential number of local optima. As it is much harder to find better basins of attraction, the correlation increases and statistics are almost the same for all stochastic hill-climber parametrizations.

Next, we evaluated how the correlation changes with the number of restarts used for mining the transition probabilities between the nodes. Figure 2 depicts

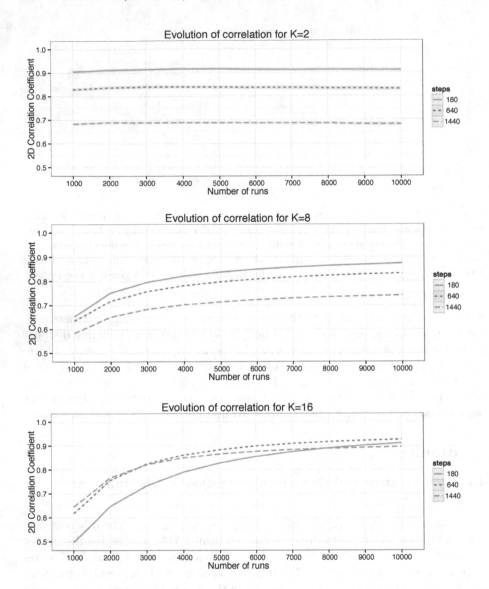

Fig. 2. Evolution of the correlation coefficient between adjacency matrices, as a function of the number of restarts of the stochastic hill-climber, for different values of K

this evolution as more restarts are used for $K = \{2, 8, 16\}$. The empirical data suggest that more restarts are needed with increasing K - diminishing landscape structure for all hill-climber parametrizations. For the most rugged NK landscape, having $K = 16$, the asymptotic increase in correlation is reached after 4000-5000 restarts. Further number of restarts increase the correlation at a diminished rate of returns.

Results of the direct comparison suggest that the data-driven approach is able to extract transition probabilities that are resembling those obtained in the literature by the deterministic operator-move based approach, without the need to systematically process the entire search space. In the followings we take a closer look on how the learnt inherent-networks capture increasing problem hardness, taking into account that we limit the size of the inherent network to at most 1024 nodes.

4.2   Structure of the Inherent Networks

Often times data visualization can provide insights into phenomena that aggregate, descriptive statistics might fail to reveal. Figure 3 shows the inherent networks that are mined from an NK landscape instance with $N = 18$ and $K = 2$ by our proposed method. The different plots present the cumulated search trajectories of Algorithm 1 that stops, respectively, after 180, 640, and 1440 *maxnoimprovement* steps. It is evident by the number and thickness of the arcs in the graphs that the more exploration allowed, the higher the density of transitions in the inherent networks. Accordingly, when the local searcher is allowed less trials without improving, it ends up more often in the same basin of departure, whereas with a larger budget of trials, it has more chances to escape weaker maxima. This can be visually appreciated from Fig. 3 considering that nodes sizes are depicted proportional to self-loop transition weights (i.e. chances of revisiting the same basin), and node color proportional to fitness: for *maxnoimprovement* $= 1440$, the rate with which some maxima are revisited has shrunk, whereas other maxima are left to dominate the search dynamics.

In order to back up such visual intuitions and to check whether the proposed method builds inherent networks whose structure correlates with the search space complexity, which for the NK landscapes is determined by the value of K, we report in Figure 4 an overview of the main network features with respect to K. Figures 4a and 4c confirm the phenomena already depicted in Fig. 3, namely that there are more connections out of node and the self-loops are less prominent when the stochastic hill-climber stops after a higher number of trials without improvement. Figures 4a and 4c present a reversal in trend after $K = 6$, which is due to the fact that the nodes of the inherent network are capped to the top 1024 local optima. For higher values of K the landscapes become more and more rugged and disordered, therefore it is harder for the local search to reach another basin of attraction (decreasing trend in average out-degree), tending to revisit the same basin (increasing trend in self-loops). The same trend reversal is exhibited by the escape edges model also, when considering only the transitions among the top 1024 local optima.

Figure 4b shows that, for all used parametrizations of the local searcher, the density of transitions steadily decreases with K, i.e. the search trajectory covers a smaller and smaller fraction of the number of possible transitions, which is expected to grow exponentially with the multimodality of the landscape. This agrees with previous findings about the local optima networks of NK landscapes [6]. The most important result is in Fig. 4d: the average length

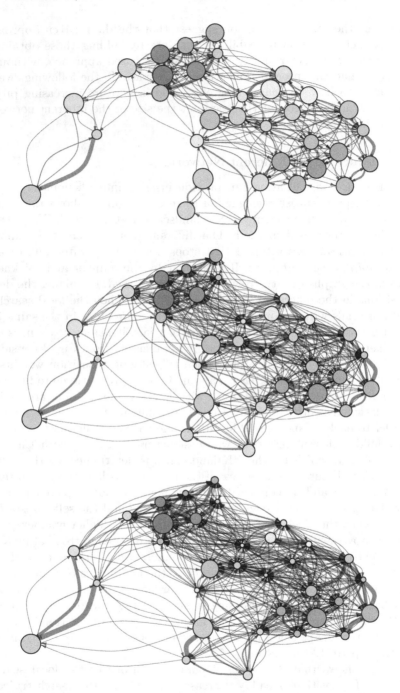

Fig. 3. Inherent networks for a typical NK instance with $N = 18$ and $K = 2$, from a stochastic hill-climber that stops after 180 (top), 640 (middle), and 1440 (bottom) steps without improvement. The size of the nodes is drawn proportional to the weight of self-loop transitions, the thickness of edges to the weight of out-going transitions. The darker the color of a node, the higher the energy value of the corresponding maximum.

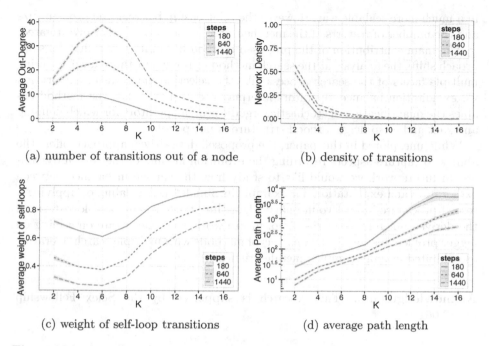

(a) number of transitions out of a node   (b) density of transitions

(c) weight of self-loop transitions   (d) average path length

Fig. 4. Main features of the inherent networks with respect to the landscape non-linearity. Lines give the average on the 100 independent instances, shaded areas (where visible) show the standard error on the mean. Plot 4d is on a semi-log scale.

of paths between local maxima, which is known to be a good predictor of problem hardness for local search heuristics, scales exponentially with the problem non-linearity, thus revealing at the inherent-network level the ruggedness of the underlying fitness landscape. This evidence confirms that, despite a limitation on the number of nodes, the proposed method can capture relevant features of the search space.

5   Conclusions

This paper proposes a data-driven approach to the off-line characterization of a combinatorial fitness landscapes by learning its inherent network from actual metaheuristic runs. Initial, classical definitions for these inherent-networks required the enumeration of the search space. Our method implies a costly compression of the input space, i.e. the training of the SOM, but such a training is to be done only once for all the instances sharing the same set and coding of solutions, and the size of the map has to be adapted to the number of nodes of the inherent network rather than to the size of the search space. Thus our results suggest that inherent networks capturing the problem instance characteristics, with the same or better descriptive power as previous methods, can be derived in a much more scalable way. However the relationship between the problem size and the number of restarts of the metaheuristic needs to be further investigated.

The main contribution of the proposed method lies in its generality. This approach shifts the analysis at the search method - problem instance level, covering multiple facets of the search process, like the selection mechanism, exploration vs. exploitation balance, operators interplay etc. In particular, the method enables the quantification of stochastic, ergodic search operators for which defining an exact and crisp neighborhood structure is not possible.

While unexplored in this paper, the proposed data-driven approach offers the ability to capture aspects regarding the evolution in time of the search dynamics. In future work we would like to study how the change in balance between exploration and exploitation, for example in Simulated Annealing or applying a reseed procedure in an Evolutionary Algorithm, reflects short and long-term in the derived inherent-networks. Also, future work will concentrate on scaling on bigger problem instances and analysis of methods with multiple search operators as exhibited in more complex metaheuristics.

Acknowledgments. This research is supported by the Sciex Fellowship nr. 12.061.

References

1. Hoos, H.H., Stützle, T.: Stochastic Local Search: Foundations and Applications. Morgan Kaufmann, San Francisco (2005)
2. Watson, J.P.: An introduction to fitness landscape analysis and cost models for local search. In: Handbook of Metaheuristics, pp. 599–623. Springer (2010)
3. Tomassini, M., Verel, S., Ochoa, G.: Complex-network analysis of combinatorial spaces: The NK landscape case. Physical Review E 78(6), 066114 (2008)
4. Verel, S., Ochoa, G., Tomassini, M.: Local optima networks of NK landscapes with neutrality. IEEE Transactions on Evolutionary Computation 15(6), 783–797 (2011)
5. Doye, J.P.K.: The network topology of a potential energy landscape: a static scale-free network. Phys. Rev. Lett. 88, 238701 (2002)
6. Vérel, S., Daolio, F., Ochoa, G., Tomassini, M.: Local optima networks with escape edges. In: Hao, J.-K., Legrand, P., Collet, P., Monmarché, N., Lutton, E., Schoenauer, M. (eds.) EA 2011. LNCS, vol. 7401, pp. 49–60. Springer, Heidelberg (2012)
7. Daolio, F., Verel, S., Ochoa, G., Tomassini, M.: Local optima networks and the performance of iterated local search. In: GECCO, pp. 369–376. ACM (2012)
8. Ochoa, G., Verel, S., Tomassini, M.: First-improvement vs. Best-improvement local optima networks of NK landscapes. In: Schaefer, R., Cotta, C., Kołodziej, J., Rudolph, G. (eds.) PPSN XI. LNCS, vol. 6238, pp. 104–113. Springer, Heidelberg (2010)
9. Kauffman, S.: The origins of order: Self organization and selection in evolution. Oxford University Press (1993)

10. Kohonen, T., Schroeder, M.R., Huang, T.S. (eds.): Self-Organizing Maps, 3rd edn. Springer-Verlag New York, Inc., Secaucus (2001)
11. Romero, G., Arenas, M.G., Castillo, P., Merelo, J.: Visualization of neural net evolution. In: Mira, J., Álvarez, J.R. (eds.) IWANN 2003. LNCS, vol. 2686, pp. 534–541. Springer, Heidelberg (2003)
12. Smith-Miles, K., Lopes, L.: Generalising algorithm performance in instance space: A timetabling case study. In: Coello, C.A.C. (ed.) LION 5. LNCS, vol. 6683, pp. 524–538. Springer, Heidelberg (2011)
13. Amor, H.B., Rettinger, A.: Intelligent exploration for genetic algorithms: using self-organizing maps in evolutionary computation. In: GECCO, pp. 1531–1538. ACM (2005)

Metaheuristics for the Pick-Up and Delivery Problem with Contracted Orders

Philip Mourdjis[1], Peter Cowling[1], and Martin Robinson[2]

[1] York Center for Complex Systems Analysis (YCCSA)
and Department of Computer Science, University of York, York, UK
{pjm515,peter.cowling}@york.ac.uk
[2] Transfaction Ltd., Suffolk, UK
martin.robinson@transfaction.com

Abstract. Contracted orders represent a novel extension to the Pick-up and Delivery Problem (PDP) with soft time windows. This extension to the multiple depot problem has depots managed by separate, competing haulage companies "carriers". Orders may be assigned to a specific carrier "contracted", "allocated" to a specific carrier but allowed to swap if this improves the solution or free to use any carrier "spot hired". Soft time windows lead to a multi-objective problem of minimising distance travelled and delay incurred. In this paper we use real order data supplied by 3 large distributors and 220 carriers. Additional, randomised, orders are generated to match the distributions observed in this data, representing backhaul orders for which no data is available. We compare a manual scheduling technique based on discussions with industry partners to popular metaheuristics for similar problems namely Tabu Search (TS), Variable Neighbourhood Search (VNS) and Hybrid Variable Neighbourhood Tabu Search (HVNTS), using our modified local search operators. Results show that VNS and HVNTS produce results which are 50% shorter than greedy approaches across test instances of 300 orders in a one week period.

1 Introduction

The purpose of this paper is to compare the effectiveness of a number of heuristic methods on a specific real world Vehicle Routing Problem (VRP), specifically a multi-depot VRP with pick up and delivery and soft time windows. Our research focusses on medium to long distance deliveries made from point to point within the UK. By considering sample orders from 3 large distributors and 220 haulage companies "carriers" we aim to reduce transportation costs and carbon emissions through the intelligent coordination of logistics activities. We are particularly interested in the gains possible through the re-assignment of orders between carriers. We note that there are currently 3 ways an order can be specified to carriers:

Contracted orders must be serviced by a specified carrier, the order may be re-allocated only between trucks belonging to this carrier.

C. Blum and G. Ochoa (Eds.): EvoCOP 2014, LNCS 8600, pp. 170–181, 2014.

Allocated   orders are assigned to specific carriers but may be re-allocated by that carrier to sub-contracted carriers.

Spot hired orders may be assigned to any truck belonging to any carrier and re-assigned any number of times to any truck.

The remainder of this paper is organised as follows. Section 2 introduces the ideas, models and concepts that we build upon in this paper. Section 3 sets out our model. Section 4 presents the local search operators used including our modification of GENI and describes the various metaheuristic methods chosen for experimentation along with the changes made to them to fit our model. Section 5 details the parameters used in our experiments, how randomised orders were generated from our existing data and compares the effectiveness of the introduced metaheuristics with varying levels of contracted and allocated orders. Section 6 presents conclusions and results analysis. Finally, we present an outline of areas for future research work.

2   Related Work

The work we have undertaken builds on VRPs with Time Windows (VRPTW), where orders must be fulfilled between a given earliest and latest time. These problems are summarised with an overview of exact algorithms and optimisation methods by Desrosiers et al. [1] and more recently with local search algorithms and metaheuristic approaches by Bräysy and Gendreau [2,3]. We also build upon research into Pick-up and Delivery Problems (PDPs) recently classified and summarised by Berbeglia et al. [4] and Parragh et al. [5]. The combination of these two areas is the PDP with Time Windows (PDPTW) [6] and still represents a lesser researched area than either of its parents. Ropke et al. [7] proposes an exact solution for the PDPTW while Malca and Semet [8] present a Tabu Search (TS) approach. Gendreau et al. [9] present neighbourhood searches for the dynamic version of this problem. For our real world problem we considered local search neighbourhoods and metaheuristics that have proved strong in the related VRPTW and tailor their methods to our specific needs. Taillard et al. [10] and Cordeau et al. [11,12] present techniques for implementing TS algorithms similar to those we use in this paper. Variable Neighbourhood Search (VNS) originally introduced by Mladenovic and Hansen [13] is a very good, general purpose, search metaheuristic capable of adapting to a wide variety of applications [14]. VNS has since been successfully applied to the VRPTW by Bräysy [15] and the multi-depot VRPTW by Polecek et al. [16]. The recent Hybrid Variable Neighbourhood Tabu Search (HVNTS) method of Belhaiza et al. [17] is tailored to the VRP with multiple time windows and was found to compare favourably to an ant colony optimisation on instances of the problem studied there.

3 Problem Definition

The PDP with contracted orders is defined on a directed graph $G = (V, A)$ where A is the arc set and $V = \{B, N\}$ is the vertex set split into B base-depot locations and N customer locations. A carrier is defined as a base location $b_i \in B$ and a set of trucks $T_i = \{T_i^0, \ldots, T_i^{M_i}\}$ where M_i is the number of trucks for carrier i. An order i consists of a collection location $c_i \in N$ and a final delivery location $d_i \in N$. In reality there are often several delivery locations as shown in Fig. 1a but at present we treat these orders as atomic, with the complexity of additional delivery locations abstracted away as in Fig. 1b for simplicity. The problem involves routing n orders into m routes, allowing for zero cost empty routes. Minimising m is not considered as part of this problem though it is kept low as a side effect of the heuristics used.

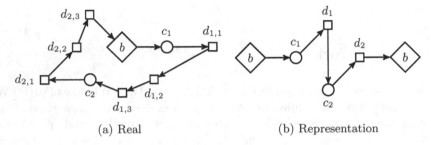

(a) Real   (b) Representation

Fig. 1. Route abstraction

Distance Model. We denote $c_{i,j}$ as the cost of travelling from order i to order j, due to the route abstraction shown in Fig. 1 these costs are asymmetric such that $c_{i,j} \neq c_{j,i}$. the straight line distance travelled while empty between the last delivery of the previous order and the current orders collection is used as the cost (Thus if a truck has no orders it has no associated cost). Each route is terminated at both ends by a dummy order located at the specified trucks base depot, thus a route j with k orders has dummy orders at 0 and $k+1$, its empty distance cost, d_j, is shown in equation 1, constrained by Max_D, the maximum distance a truck is permitted to drive in a week.

$$d_j = \sum_{i=0}^{k} c_{i,i+1}.\text{where } d_j < \text{Max}_D \tag{1}$$

Any change to the solution can be mapped to a series of insertion and removal operations. As the orders themselves are present in both solutions (before and after any change) the only aspects that need to be considered are the legs between orders, shown in Fig. 2. Denoting $\check{c}_i$ as the insertion cost of an order i between two pre-existing orders x and y, the change caused by inserting an order is calculated as shown in equation 2.

$$\check{c}_i = c_{x,i} + c_{i,y} - c_{x,y}. \tag{2}$$

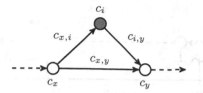

Fig. 2. Route alteration

Similarly the removal cost $\hat{c}_i$ of an order is as shown in equation 3:

$$\hat{c}_i = c_{x,y} - c_{x,i} - c_{i,y}. \tag{3}$$

A key point to note is that both $\hat{c}_i$ and $\check{c}_i$ may be positive or negative with positive costs indicating an increase in empty distance and negative costs indicating a decrease.

3.1 Time Window Model

Figure 3 shows a number of collection time windows. e_i is the earliest time a truck may service customer c_i and l_i is the latest time, there is no penalty for arriving at a location early, though the truck will have to wait until the specified earliest time to be serviced. If the truck arrives after l_i the order is said to be delayed by $t_i - l_i$ where t_i is the actual time customer i is serviced. Not shown in Fig. 3 is the service time required for loading / unloading at a customer location, this is denoted by s_i. Tardiness is calculated based on the vehicles arrival time at a location thus if a vehicle arrives at the latest arrival time the tardiness is 0 even though the truck will not leave until $s_i + l_i$ (after the latest time l_i). The tardiness of a vehicle t_V is simply the sum of the individual delays experienced at each location in its route. Orders are always inserted as early as possible at the chosen insertion point and changes to a route force an update of delay parameters for each subsequent location.

Note, that since a collection node must occur before its delivery node, reversing a section of a route will significantly alter the distance, time windows are also usually tight enough such that one or more orders will be rendered significantly delayed. Methods relying on partial route inversions such as GENI [18] and iCROSS [15] will therefore not work well without alteration.

3.2 Objective

Our optimisation procedure seeks to fulfil all orders in such a way as to reduce the total travel cost whilst keeping tardiness to a minimum. The fitness of a vehicles route, f_V, is given in equation 4. Here α represents a tunable parameter between 0 and 1 determining the relative importance of tardiness and distance respectively. D_V and T_V represent the total distance and time of a vehicles route

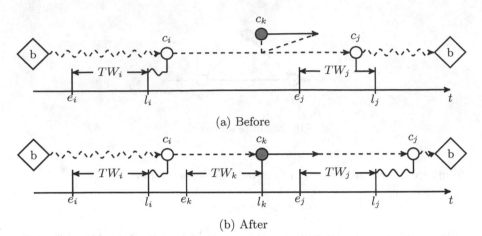

(a) Before

(b) After

Fig. 3. Tardiness - Two orders c_i and c_j are currently scheduled and only c_i is delayed. If another order (c_k) were placed between the two existing orders, c_i would remain at its current time while c_j may have to occur later, potentially becoming delayed.

respectively, dividing by these gives relative empty miles and relative tardiness, allowing comparisons to be made between the two metrics which would otherwise be orders of magnitude different.

$$f_V = \alpha \left(\frac{d_V}{D_V} \right) + (1 - \alpha) \left(\frac{t_V}{T_V} \right) \tag{4}$$

Since the impact of both of these upon a carrier is in additional cost (or lost profits) we combine them into a single objective function. α therefore determines the relative cost of driving additional miles versus late delivery penalties. The fitness of a solution is the sum of the individual fitnesses of all its vehicles as shown in equation 5.

$$F_S = \sum_{j=1}^{m} f_j. \tag{5}$$

4   Solution Methods

4.1   Local Search Operators

For hundreds of carriers and thousands of orders it is computationally intensive to calculate a fitness from a solution. We use local moves to make incremental changes to the current solution instead and measure changes in fitness. These are much easier to compute and over successive iterations we can make large changes to the solution.

A number of local search operators are used including cross [10], relocate [2], swap [2] and a modification of GENI [18]. Each of these local moves is intended

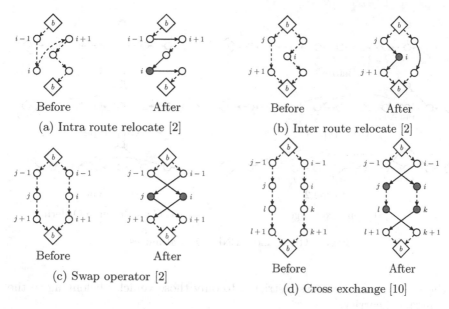

(a) Intra route relocate [2]   (b) Inter route relocate [2]

(c) Swap operator [2]

Before   After

(d) Cross exchange [10]

Fig. 4. Local moves

to preserve existing orderings as much as possible. Our modification of GENI is presented below and the other operators used are summarised in Fig. 4. They represent restricted 3- and 4-opt operators which preserve the order of nodes.

GENI - Preserve Ordering. A local move, similar in spirit to the generalized insertion (GENI) procedure of Gendreau et al. [18] was devised as follows, for a given order to be inserted into a chosen target route, for each pair of nodes in the target route, calculate the two insertion costs as shown in Fig. 5b using equation 2. In comparison to GENI, Fig. 5a, GENI-PO does not reverse the traversal of any existing arcs of the solution and should be more suitable for this real world problem with time windows.

4.2 Metaheuristics

We sought to make comparisons between popular and contemporary metaheuristics from the literature and a greedy assignment without optimisation. We use simple versions of TS [19,10], VNS [13,14] and HVNTS [17] for our experiments, adapted such that they are effective for our problem and a fair comparison can be made between them. To this end a number of differences to the original methods have been made.

1. We have adapted each procedure to use the same set of local search operators, namely intra- and inter-route relocate, swap, cross and GENI-PO introduced in Section 4.1. When optimising routes, we check if the order we want to move is contracted, if it is, we can still move the order between vehicles but

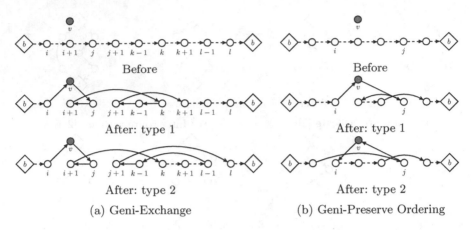

Before   Before

After: type 1   After: type 1

After: type 2   After: type 2

(a) Geni-Exchange   (b) Geni-Preserve Ordering

Fig. 5. GENI and GENI-PO local moves

the search space becomes restricted to only those vehicles belonging to the contracted carrier.

2. In each case a greedy insertion method is used to generate an initial solution. This method takes a random list of orders and inserts each in its lowest cost insertion location given already scheduled orders. For each insertion location, the lowest insertion cost $min\,(\check{c}_i)$ is determined by using equation 2 for each valid insertion point. The greedy insertion method is used as a baseline for the comparison after discussions with our sponsor, Transfaction, as a technique which closely mimics current manual/semi-automatic approaches to scheduling.

3. Once an initial solution has been generated the three methods are each given 50 CPU seconds to generate a result, allowing fast iterating techniques to run more iterations. As the heuristics use differing amounts of CPU time to process one iteration comparing the heuristics with a fixed number of iterations would not be fair. In the extreme a heuristic approach may be "beaten" by an exhaustive style search if the same iterations of each were performed but would take substantially (thousands of times) longer to run. We feel that keeping CPU time constant is a fair way to evaluate these methods [20]. All tests were carried out on a Windows 7 SP1 desktop machine running C$^\sharp$ code on a 2.8Ghz Intel core i5-2300 processor with 8Gb of RAM.

5   Computational Experiments

5.1   Generating Orders

At present we have access to one week's worth of order data for 3 large distributors and 220 hauliers. To more thoroughly test our heuristics and to ensure we are not overfitting to our sample data we generate additional randomised orders in the form of backhauls [4]. These are derived from existing orders where collection locations are picked from real delivery locations and delivery locations are picked from real collection locations[1]. We use a set of uniform and Poisson distributions with parameters tuned to approximate the orders for which we have data.

5.2 Speed and Travel Parameters

Each collection and delivery location used is based on UK postcodes which are translated to standard eastings and northings. Between locations, straight line distances are used and it is assumed that trucks travel at a constant 35 Kph as this is the average value derived from our massive data set. We set Max_D at 1650 km as at 35 Kph this is the limit for the number of hours a long haul driver is allowed to drive in a week.

5.3 Aims

To investigate the effect of contracted and allocated orders, we conducted 5 sets of experiments, in each set, orders were defined with: 100% contracted; 30% contracted, 60% allocated & 10% spot hired; 30% contracted & 70% spot hired; 60% allocated & 40% spot hired and all spot hired respectively. In each case 10 seeded randomised runs were performed for each heuristic. For each run, 200 real orders were selected from a database of orders and a further 100 were generated as described in Section 5.1.

5.4 Findings

To easily display the large numbers of results generated, groups of four box plots have been used to represent the four techniques compared. Each box plot represents the min / max and quartiles of the 10 runs. Figure 6 presents this information along with the order of the heuristics used in the following charts. Here "Greedy" represents the initial solution before optimisation, VNS, TS and HVNTS represent the results from our modified heuristics after optimisation. Figures 7 and 8 present the empty miles and delay for all heuristics across the range of scenarios introduced above. We can see that there is a clear trend towards shorter distances as we allow orders more freedom in carrier choice. This trend is amplified by our metaheuristics which produce little benefit in a fully contracted model but produce benefits of approximately 50% in the spot model.

Of the three heuristic approaches investigated, VNS and HVNTS can be seen to produce the shortest routes across multiple runs of our experiments when orders are spot hired. When all orders are contracted to a specific carrier there is much less variation in the results observed, note that there is still a large

---

[1] We know backhaul orders of this kind exist but have no data for them, since they are often requested by small distributors.

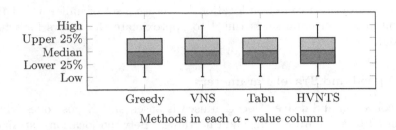

Fig. 6. Key

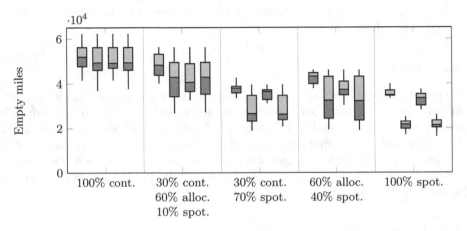

Fig. 7. Empty miles - lower is better

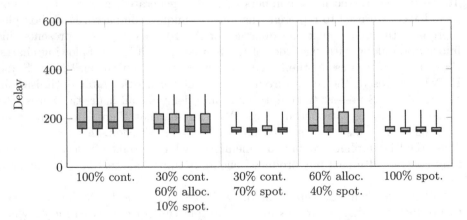

Fig. 8. Delay - lower is better

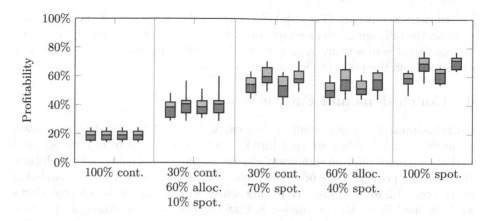

Fig. 9. Average carrier profitability - higher is better

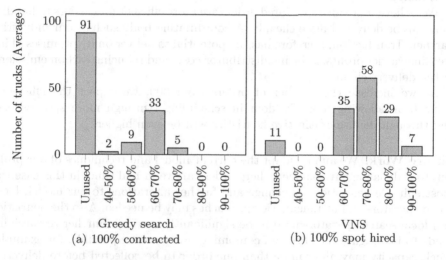

Fig. 10. Route profitability breakdown

solution set to evaluate in this case as even when contracted to a carrier there remains a choice of delivery vehicle and ordering. We feel that due to time order constraints the greedy insertion heuristic used in the entirely contracted examples is able to produce routes that are close to optimal and do not leave room for our heuristic techniques to improve upon.

Figure 9 shows that the average profit a carrier attains under any heuristic increases as the proportion of spot hired orders increases. Here profitability is the percentage of distance travelled that is spent on delivery, between pick-up and delivery.

We also observed that the fully contracted scenario using greedy scheduling produced an unfair distribution of orders between carriers such that many were left without any orders (Fig. 10a). Moving to the other end of the spectrum, VNS on the fully spot hired scenario produces higher rates of profitability which are consistent across many more carriers (Fig. 10b), yielding a more sustainable situation from the point of view of the carriers.

6   Conclusions and Future Work

Conclusions. From our results it is clear to see that the shortest routes are achieved when all orders are spot hired, free to be assigned to any carrier, and that these routes offer no significant change in the overall delay of the solution. We note that pre-allocating orders to preferred carriers, though better than being contracted, still produces relatively long routes with the optimisation procedures we have used here. Also of interest is that our optimisation strategies produce far greater benefits over the initial solution when there are no contracted or allocated orders. This can likely be attributed to the larger solution space available to explore.

We believe that since spot hired orders are more efficient for carriers to handle, they can be delivered more cheaply. A coordinating body such as our industrial partner, Transfaction, therefore has the potential to deliver on its promises of increasing carrier profits, reducing distributor costs and reducing carbon emissions in the delivery chain.

As we increase the number of orders from 300, we expect the efficiency gains from allowing more freedom in scheduling (through more spot orders) and through the use of effective heuristics will be even higher.

Future Work. We aim to adapt the current model and techniques of our problem to a dynamic environment where orders arrive in real time, in this case the most suitable heuristic may change and further alterations to our model, local move operators and metaheuristic approaches may be needed. Also the heuristics and local search operators need to be significantly sped up. Further research intends to investigate the effects of combining orders where possible, for example, truck capacity may allow more than one order to be collected before deliveries commence.

Acknowledgements. This work has been funded by the Large Scale Complex IT Systems (LSCITS) EngD EPSRC initiative in the Department of Computer Science at the University of York and by Transfaction Ltd.

References

1. Desrosiers, J., Dumas, Y., Solomon, M.M., Soumis, F.: Time Constrained Routing and Scheduling. Handbooks in Operations Research and Management Science 8, 35–139 (1995)
2. Bräysy, O., Gendreau, M.: Vehicle Routing Problem with Time Windows, Part I: Route Construction and Local Search Algorithms. Transportation Science 39(1), 104–118 (2005)

3. Bräysy, O., Gendreau, M.: Vehicle Routing Problem with Time Windows, Part II: Metaheuristics. Transportation Science 39(1), 119–139 (2005)
4. Berbeglia, G., Cordeau, J.F., Gribkovskaia, I., Laporte, G.: Static pickup and delivery problems: a classification scheme and survey. TOP 15(1), 1–31 (2007)
5. Parragh, S.N., Doerner, K.F., Hartl, R.F.: A survey on pickup and delivery problems. Journal für Betriebswirtschaft 58(2), 81–117 (2008)
6. Dumas, Y., Desrosiers, J., Soumis, F.: The pickup and delivery problem with time windows. European Journal of Operational Research 54, 7–22 (1991)
7. Ropke, S., Pisinger, D.: An Adaptive Large Neighborhood Search Heuristic for the Pickup and Delivery Problem with Time Windows. Transportation Science 40(4), 455–472 (2005)
8. Malca, F., Semet, F.: A tabu search heuristic for the pickup and delivery problem with time windows and a fixed size fleet, 1–5 (2003) (unpublished manuscript)
9. Gendreau, M., Guertin, F., Potvin, J.Y., Séguin, R.: Neighborhood Search Heuristics for a Dynamic Vehicle Dispatching Problem with Pick-ups and Deliveries. Transportation Research Part C: Emerging Technologies 14(3), 157–174 (2006)
10. Taillard, E.D., Badeau, P., Gendreau, M., Guertin, F., Potvin, J.Y.: A Tabu Search Heuristic for the Vehicle Routing Problem with Soft Time Windows. Transportation Science 31(2), 170–186 (1997)
11. Cordeau, J.F., Laporte, G., Mercier, A.: A unified tabu search heuristic for vehicle routing problems with time windows. Journal of the Operational Research Society 52(8), 928–936 (2001)
12. Cordeau, J.F., Laporte, G., Mercier, A.: Improved tabu search algorithm for the handling of route duration constraints in vehicle routing problems with time windows. Journal of the Operational Research Society 55(5), 542–546 (2004)
13. Mladenović, N., Hansen, P.: Variable Neighbourhood Search. Computers & Operations Research 24(1), 1097–1100 (1997)
14. Hansen, P., Mladenović, N., Moreno Pérez, J.A.: Variable Neighbourhood Search: Methods and Applications. Annals of Operations Research 175(1), 367–407 (2009)
15. Bräysy, O.: A Reactive Variable Neighborhood Search for the Vehicle-Routing Problem with Time Windows. INFORMS Journal On Computing 15(4), 347–368 (2003)
16. Polacek, M., Hartl, R.F., Doerner, K.F.: A Variable Neighborhood Search for the Multi Depot Vehicle Routing Problem with Time Windows. Journal of Heuristics 10, 613–627 (2004)
17. Belhaiza, S., Hansen, P., Laporte, G.: A hybrid variable neighborhood tabu search heuristic for the vehicle routing problem with multiple time windows. Computers & Operations Research (August 2013)
18. Gendreau, M., Hertz, A., Laporte, G.: New Insertion and Post Optimization Procedures for the Traveling Salesman Problem. Operations Research 40(6), 1086–1095 (1992)
19. Glover, F.: Artificial Intelligence, Heuristic Frameworks and Tabu Search. Managerial and Decision Economics 11(5), 365–375 (1990)
20. Barr, R.S., Golden, B.L., Kelly, J.P., Resende, M.G.C., Stewart, W.R.: Designing and Reporting on Computational Experiments with Heuristic Methods. Journal of Heuristics 1(1), 9–32 (1995)

Modeling an Artificial Bee Colony with Inspector for Clustering Tasks

Cosimo Birtolo, Giovanni Capasso, Davide Ronca, and Gennaro Sorrentino

Poste Italiane – Information Technology,
S - FSTI - R&D Center – Piazza Matteotti 3 – 80133 Naples, Italy
{birtoloc,capass56,roncadav,sorre137}@posteitaliane.it

Abstract. Artificial Bee Colony (ABC) is a recent meta-heuristic approach. In this paper we face the problem of clustering by ABC and we model a further bee role in the colony, performed by inspector bee. This model conforms with real honey bee colony, indeed, in nature some bees among the foraging ones are called inspectors because they preserve the colony's history and historical information related to food sources. We experiment inspector behavior in ABC and compare the solution to traditional clustering algorithm. Finally, the effect of colony size is investigated and experimental results are discussed.

Keywords: Artificial Bee Colony, Soft Computing, Clustering, Inspector, Data Mining.

1 Introduction

Clustering algorithms play a relevant role in understanding and exploring a dataset. Interest in clustering algorithm is proved by the need of knowledge extraction processes in a huge amount of data in several domains: from bioinformatics to web usage mining, from image segmentation to information retrieval.

Clustering aims at minimizing the dissimilarity between data assigned to the same cluster and is a powerful tool which can arise interesting information in the area it is applied. Moreover, clustering can be considered one of the most difficult and challenging problems in machine learning, particulary due to its unsupervised nature.

Clustering problems have been solved using various techniques, even if K-means, independently discovered in different scientific studies in the 60's [1], is one of the most popular algorithms due to its simplicity and efficiency. However, the clusters resulting from the K-means algorithm are very sensitive to positions of the initial centroids in the problem space and the algorithm can converge to a local optimum. Recently, meta-heuristic approaches have been proposed to solve clustering problems [2] and in particular some Artificial Bee Colony have been adopted for this task [3].

In this paper, we experiment ABC approach to clustering tasks by means of different datasets. The merits of this contribution are: (i) the introduction and the modeling of a new role in the colony, i.e., the inspector bee in an ABC

C. Blum and G. Ochoa (Eds.): EvoCOP 2014, LNCS 8600, pp. 182–193, 2014.

algorithm, and (ii) the experimentation on the way the colony size and its composition can influence the algorithm's results.

The remainder of this paper is organized as follows: Section 2 introduces the clustering; Section 3 describes Artificial Bee Colony (ABC); Section 4 depicts the proposed formulation of ABC; Section 5 describes the algorithm structure, Section 6 provides experimental results; and Section 7 outlines conclusions and future directions.

2   The Clustering Problem

Clustering algorithms aim at grouping data into a number of clusters. Data in the same cluster share a high degree of similarity while they are very dissimilar from data of other clusters. Partitional clustering algorithms aim at partitioning the population into a fixed number k of classes, each of those being represented by an average item named centroid. The traditional partitional clustering algorithm is K-means [1] which has been applied to a wide range of problems in different domains. However, K-means is sensitive to the initial states and can converge to the local optimum solution. Recently, many methods have been proposed in order to overcome this drawback.

The clustering problem can be stated as the minimization of the sum of Euclidean squared distance between each object x_i and the center of the cluster c_j to which it belongs (i.e., centroids). The objective function to be minimized can be expressed by Eq.1:

$$J(w,c) = \sum_{i=1}^{N}\sum_{j=1}^{K}\sum_{d=1}^{D} w_{ij} \left\| x_{i,d} - c_{j,d} \right\|^2 \tag{1}$$

where K is the number of clusters, N is the number of objects, w_{ij} is the association weight of objects x_i in cluster j, i.e., $w_{i,j}$ is 1 if object i is allocated to cluster j, and 0 if it is not. Each data instance x_i and each cluster center c_j is defined by a vector of D values, where D is the number of features. The center of each of the j cluster $c_j = \{c_{j,1}, c_{j,2}, \ldots, c_{j,D}\}$ is the set of the mean of each dimension across all the objects assigned to the jth cluster and it can be calculated by Eq.2 below

$$c_{j,d} = \frac{1}{N_j}\sum_{i=1}^{N} w_{i,j} x_{i,d} \tag{2}$$

where N_j is the number of objects in the jth cluster.

Different evolutionary approaches are adopted to address clustering tasks (i.e., fixed or variable number of clusters, centroid-based, medoid-based, label-based, tree-based or graph-based representation) as described by Hruschka et al.[2], but recently some swarm intelligence techniques are proposed [3,4].

3   Artificial Bee Colony

Artificial Bee Colony (ABC) Algorithm is a recent swarm intelligence algorithm based on the intelligent behavior of honey bee foraging. It was proposed by Karaboga [5] in 2005 and performances are analyzed in 2008 [6]. ABC is based on modeling the behaviors of real bees on finding nectar amounts and sharing the information of food sources to the other bees in the hive.

Honey bees are social insects and live in large organized communities. Each bee has specific skills and carries out determined works with the aim of facilitating the survival of the colony. The provision of the food is one of the major activities within a colony. This activity involves specific worker bees which collaborate among each other: the "employed bees", which research and communicate where the food sources are; the "onlooker bees" which extract and carry the food. The main task of an employed bee is to look for food. When the food source has been found, the bee memorizes the spatial coordinates and communicates the position and the quality of the source through a dance around the hive. The dance and the research activity alternate each other. The main task of an onlooker bee is observing the employed bees dance outside the hive. On the basis of the message expressed by the dance, the onlooker bee chooses the food source that best fits its needs. After the choice, the onlooker bee reaches the source to extract the food and carry it to the hive. On the way to the food source, the onlooker bee may discover a better food source than the chosen one. In this case, when the onlooker bee goes back to the hive communicates the position of the new food source to the employed bee. When a food source is finished by onlooker bees, the employed bee, that was communicating that source's position, forgets those coordinates and looks for a new source.

Taking inspiration from the nature, Karaboga models three bee behaviors in the colony: (i) The Employed Bee, (ii) The Onlooker Bee, and (iii) The Scout Bee. The employed bees are associated with the specific food sources, onlooker bees watch the dance of employed bees within the hive to choose a food source, and scout bees look for food sources randomly [5].

In nature, the employed bee whose food source has been exhausted becomes a scout bee to look for the further food sources. In ABC, the solutions represent the food sources and the nectar quantity of the food sources corresponds to the fitness of the associated solution. Employed bees whose solutions was not improved after a fixed number of trials, defined *limit*, become scouts and their solutions are abandoned.

In other words, the general formulation of the ABC algorithm can be described by the following phases: (i) Bee Initialization, (ii) Employed Bee Phase (iii) Onlooker Bee Phase (iv) Scout Bee Phase (v) Memorization of the best solution found. These last four phases are iterated until the stop criteria is met. Commonly the algorithm stops when a fixed maximum number of cycles is reached.

Nowadays, different real world applications of ABC algorithm [7] have been investigated. In 2011, Karaboga and Ozturk [3] firstly introduced ABC for clustering tasks, showing how ABC formulation outperformed Particle Swarm Optimization (PSO) algorithm. Moreover the authors experimented ABC in classification tasks, comparing it with traditional classification algorithms such as Neural Networks (Multi Layer Perceptron), Bayesian Network, Radial Basis Function (RBF) proving the benefits of bee colony. A first hybrid approach is proposed by Yan et al. [4] who present a Hybrid Artificial Bee Colony algorithm. The authors consider a social learning between bees by means of cross-over operators of Genetic Algorithm and apply the proposed algorithm to some classification tasks proving some benefits despite of traditional k-means, ABC and PSO algorithm.

4   Inspector Bee in the Colony

Our proposed algorithm is inspired by the Simple ABC given by Karaboga [6], but it extends the colony modeling a forth bee behavior, i.e., Inspector Bee.

In a real bee colony, inspection role was modeled by Biesmeijer and de Vries [8], who introduced additional behavioral states for forager bees. The authors defined 7 different bee behaviors and the transitions between them: (i) novice forager, (ii) scout, (iii) recruit, (iv) employed forager, (v) unemployed experienced forager, (vi) inspector, and (vii) reactivated forager. In their work they define the inspectors as foragers that retire from an unprofitable food source but continue to make occasional trips to it, while reactivated foragers are bees that stop inspecting after a certain period of time and return to wait for dances to follow at the nest.

Granovskiy et al. [9] studied the role of inspector bees. Their experiments show that a bee colony is able to successfully reallocate its foraging resources in dynamic environments even when dance language information is limited. According to the authors, it remains unclear in what foraging situations reactivation and inspection are important and in what cases the dance language is the primary mechanism for communicating memory. The ability of the colony to react to rapid changes in their environment can be justified by the inspector bees that act as the colony's short-term memory [8]. So that, these bees allow the colony to quickly begin utilizing previously abandoned food sources once they become profitable again.

Inspection can be considered an important mechanism for reallocating foragers when food sources are hard to find: for these reasons we introduce inspector in the proposed Artificial Bee Colony. In our model, the Inspector Bee memorizes the best solution across the different cycles, so that if a solution is abandoned by bees and is not considered as the best solution for the next cycle, the inspector preserves this information.

5   Algorithm Structure and Fitness Function

Pseudo-code of our Artificial Bee Colony with Inspector behavior (ABCi) is outlined by Algorithm 1. The parameters of the proposed ABC algorithm as well as Karaboga's formulation are: the number of food sources (i.e., SN), the

---

Algorithm 1. ABCi: algorithm's pseudo-code

1: Load training samples
2: Set the number of employed bees and onlooker bees
3: Generate the initial population z_s, $s = 1..SN$ with trial counter $t_s = 0$
4: Evaluate the nectar amount (fitness function) of the food sources ($\forall s$)
5: Inspector bee moves to the best food source
6: Set cycle to 1
7: **repeat**
8:      **for all** employed bee assigned to solution s **do**
9:          Produce new solution v_s with $t_s = 0$
10:          Evaluate the fitness of the new solution v_s
11:          Apply greedy selection process for the identification of new population z_s
12:      **end for**
13:      Calculate the probability values p_s for the solutions z_s, $s = 1..SN$
14:      **for all** onlooker bee **do**
15:          Select a solution z_s depending on p_s
16:          Produce new solution v_s with $t_s = 0$
17:          Evaluate the fitness of the new solution v_s
18:          Apply greedy selection process for the identification of new population z_s
19:          **if** greedy selection process preserves old solution **then**
20:              Increment the trial counter t_s associated to the solution z_s
21:          **end if**
22:      **end for**
23:      Inspector bee moves to the best food source and memorize it
24:      **if** there is a solution with $t > limit$ (scout bee) **then**
25:          Generate a new solution according a randomized process
26:          Memorize the new solution, replacing the abandoned one
27:      **end if**
28:      $cycle = cycle + 1$
29: **until** cycle = MCN

---

number of employed and onlooker bees, the value of the *limit*, and the maximum cycle number (MCN).

In clustering problem the food sources are the cluster centroids, while the solution is the position of food source which maximizes the nectar amount (the position of centroids which minimizes the fitness function).

In the initialization phase, the algorithm generates randomly a group of food sources corresponding to the solutions in the search space. According to Eq.3, the fitness of food sources is evaluated and for each food source a counter which stores the number of trials of each bee is set to 0 in this phase.

$$fitness(s) = \sum_{i=1}^{N} \sum_{j=1}^{K} w_{i,j} \left\| x_i - c_j \right\|^2 \tag{3}$$

where K is the number of clusters, N is the number of objects, x_i is a generic input to be clustered, c_j is the jth centroid, and s is the solution (the position of K centroids).

In the employed bees' phase (see lines 8-13 in algorithm's pseudo-code), each employed bee is sent to the food source and finds a neighboring food source. The neighboring food source is produced according to Eq.4 as follows:

$$v_{i,j} = z_{i,j} + \phi\left(z_{i,j} - z_{k,j}\right) \tag{4}$$

where k is a randomly selected food source different from i, j is a randomly chosen centroid. ϕ is a random number between [-1,1]. The new food source v is determined by changing randomly one dimension on jth centroid. If the produced value exceeds its predetermined boundary, it will set to be equal to the boundary. Then the new food source is evaluated. Therefore, a greedy selection is applied. In other words, the employed bee produces a modification in the position (i.e. solution) and checks the nectar amount (fitness value) of that source (solution). The employed bee evaluates this nectar information (fitness value) and then assigns to the food source a probability related to its fitness value according to the Eq.5.

$$p(s) = f(s) \left/ \sum_{j=1}^{K} f_j \right. \tag{5}$$

where K is the number of food sources and $f(s) = \frac{1}{1+fitness(s)}$

In the onlooker bees' phase (see lines 14-23 in algorithm's pseudo-code), the onlooker bee selects a food source based on a probability of a source explored by employed bees. Once the food sources have been selected, each onlooker bee finds a new food source similarly to the employed bee (see Eq.4) and the greedy selection process select the new source. If this process preserves old solution, the value of counter, which is associated to the employed bee, increases.

In scout bees' phase (see lines 24-27 in algorithm's pseudo-code), when the value of the counter t of a food source is greater than $limit$, the food source is abandoned, the inspector bee memorizes the source and the employed bee becomes a scout. The scout bee generates a new solution according to Eq.6 and sets the value of counter equal to 0, so that the bee memorizes the new solution replacing the abandoned one.

$$z_{j,d} = \min_{i=1}^{N}\left(x_{i,d}\right) + rand(0,1) \cdot \left(\max_{i=1}^{N}\left(x_{i,d}\right) - \min_{i=1}^{N}\left(x_{i,d}\right)\right) \tag{6}$$

where $j = 1, 2, ...K$ and $d = 1, 2, ..., D$. N is the number of objects, K is the number of clusters, and D is the number of features. $x_{i,d}$ represents the d-th feature of the input data x_i.

6   Experimental Results

In this section we experiment the proposed ABC algorithm for some clustering problems and for an application in Transportation System.

In order to evaluate the performance of the proposed ABC approach, we compare the results of the K-means, ABC, and the proposed ABC for a clustering task by comparing four different datasets. These datasets are selected from the UCI machine learning repository (Breast Cancer Wisconsis, Credit Approval, Dermatology and Iris datasets) [10]. An additional dataset, which have been extracted from a real-world clustering problem in Poste Italiane domain, is considered as an example of application.

Iris data was collected by Anderson in 1935 and consists of 150 random samples of flowers from the iris species setosa, versicolor, and virginica. From each species there are 50 observations for sepal length, sepal width, petal length, and petal width in cm.

Wisconsin Breast Cancer consists of 683 objects characterized by 9 features: clump thickness, cell size uniformity, cell shape uniformity, marginal adhesion, single epithelial cell size, bare nuclei, bland chromatin, normal nucleoli, and mitoses. There are two categories in the data: malignant (444 objects) and benign (239 objects).

Credit Approval dataset contains 690 samples, which are different credit card applications, with 15 attributes. This dataset has a good mix of attributes (continuous, nominal with small numbers of values, and nominal with larger numbers of values) and data can be grouped either in approved or not approved.

Dermatology consists of 366 samples characterized by 34 features which are 12 patient clinical attributes and 22 histopathological features. The values of the histopathological features are determined by an analysis of the samples under a microscope. The diseases in this group can be one of the following six: psoriasis, seboreic dermatitis, lichen planus, pityriasis rosea, cronic dermatitis, and pityriasis rubra pilaris.

6.1   Convergence Analysis

We run the algorithm several times with different value of *limit* in order to study quantitatively the convergence of the two different ABC formulations. We consider the four datasets from UCI database as benchmark data.

The parameters in an ABC approach are the *limit* and the colony size [11], so that we study algorithm's performance as long as the parameters change.

We repeated 20 runs for different problem configurations. First of all, cycle after cycle, we report the average of best fitness with different abandonment behavior of a nectar source (*limit* is equals to 0, 5, 10, 20, 50, 100, 1000) when a colony of 20 bees is considered.

To find a better solution, one may search the largely unknown region (*exploration*) or search around the current solution (*exploitation*). The tradeoff between exploration and exploitation is represented by the *limit*. Indeed, higher values of limit emphasize exploitation behavior of the algorithm, while lower values of limit foster exploration phase.

Best solutions occurs when *limit* increases, as the exploitation behavior becomes more relevant. On the other hand, very high value (i.e., 1000) of limit holds algorithm back for exploration of new solutions. However, we can notice

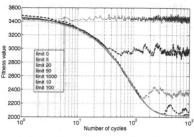
(a) Inspector is not considered

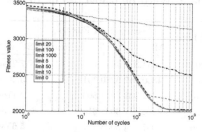
(b) Inspector is considered

Fig. 1. Dermatology dataset: Average fitness behavior by varying the *limit*. Colony size equals 110.

Table 1. Wilcoxon paired test on Iris dataset: Average fitness and p-values

Limit	Average Best Fitness			p-value		
	ABC	ABCi	k-Means	ABC vs. ABCi	ABC vs. k-Means	ABCi vs. k-Means
0	214.278	140.450		7.254e-12	7.562e-09	7.562e-09
5	139.540	101.306		7.254e-12	9.637e-08	3.016e-06
10	111.508	97.461	99.990	1.451e-11	1.006e-06	1.709e-01
20	97.956	96.698		3.685e-09	1.709e-01	1
50	96.675	96.656		1.066e-01	1	1
100	96.659	96.655		1.024e-02	1	1
1000	96.655	96.655		8.318e-01	1	1

how ABCi's convergence is not heavily affected by limit value if they ranges between 20 and 100, thus resulting algorithm to be robust to this situation.

Furthermore, *limit* equals to 50 (black curve) could be a good tradeoff, even if the optimal parameter value depends on the particular problem. Indeed, the Dermatology dataset (see Fig.3) seems not to converge with *limit* equal to 0, 5 and 1000. Moreover this dataset presents high convergence time with other values. The problem is the colony size which must be incremented as proved in Fig.1 when we repeat the 20 different runs with 100 onlookers and 10 employers.

Investigating these results more deeply, we consider Mann-Whitney-Wilcoxon test, reporting results in Table 1, where the average value of best fitness of 20 different trials per technique (i.e., ABC, ABC with inspector, k-means). The null hypothesis is: the investigated techniques provide solutions which belong to the same population entailing a comparable clustering performance, and the alternative hypothesis is: (i) ABC fitness is greater than ABCi one, (ii) ABC fitness is greater than k-means, and (iii) ABCi fitness is greater than k-means.

Assuming 0.05 as upper bound to reject the null hypothesis, we can affirm that there is statistical difference between ABC and ABCi. We prove that ABCi outperforms ABC because ABCi provides a lower fitness value in most of the cases. We cannot reject the null hypothesis with higher value of limit (i.e., *limit*

Table 2. Wilcoxon paired test on Breast Cancer dataset: Average fitness and p-values

Limit	Average Best Fitness			p-value		
	ABC	ABCi	k-Means	ABC vs. ABCi	ABC vs. k-Means	ABCi vs. k-Means
0	7254.312	4645.158		7.254e-12	4.003e-09	4.003e-09
5	4860.267	3280.301		3.265e-10	4.003e-09	4.003e-09
10	3462.931	3044.500	3061.098	2.176e-10	1.996e-06	1
20	3165.311	3035.615		1.183e-04	1	1
50	3037.424	3035.571		5.658e-01	1	1
100	3035.571	3035.571		8.367e-01	1	1
1000	3035.571	3035.571		3.088e-01	1	1

Table 3. Wilcoxon paired test on Credit Approval dataset: Average fitness and p-values

Limit	Average Best Fitness			p-value		
	ABC	ABCi	k-Means	ABC vs. ABCi	ABC vs. k-Means	ABCi vs. k-Means
0	4.399e+06	5.958+05		7.254e-12	4.003e-09	1
5	1.617e+06	5.731e+05		8.705e-11	1,267e-01	1
10	6.183e+05	5.624e+05	8.087e+05	7.254e-12	1	1
20	5.614e+05	5.571e+05		6.673e-06	1	1
50	5.570e+05	5.568e+05		9.328e-01	1	1
100	5.570e+05	5.568e+05		7.858e-02	1	1
1000	5.568e+05	5.568e+05		9.214e-01	1	1

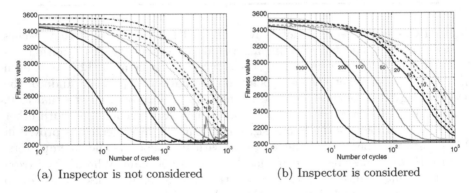

(a) Inspector is not considered   (b) Inspector is considered

Fig. 2. Dermatology dataset: Average fitness behavior by varying the number of on-lookers (10 employers)

greater than 100) and ABC and ABCi performance are comparable. Indeed, considering a higher value of limit, the abandonment behavior of an employed bee decreases and the benefit of an inspector bee is not estimable.

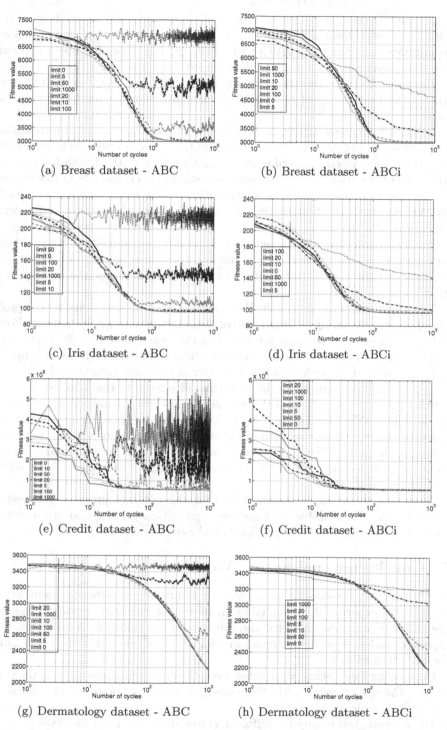

(a) Breast dataset - ABC

(b) Breast dataset - ABCi

(c) Iris dataset - ABC

(d) Iris dataset - ABCi

(e) Credit dataset - ABC

(f) Credit dataset - ABCi

(g) Dermatology dataset - ABC

(h) Dermatology dataset - ABCi

Fig. 3. Average fitness behavior by varying the *limit*. Colony size equals to 20.

Instead, comparing k-means with ABC approach, we prove how a bee colony can outperform with *limit* greater than 20. Low values of limit penalizes exploitation behavior, while ABCi with *limit* equal to 20 or 50 is able to provide promising results in clustering problem, improving k-means results.

The same findings arise from Breast Cancer dataset (e.g., see Tab.2), confirming the importance of *limit* and the benefits of inspector bee within the colony.

Taking into account Credit Approval dataset (see Tab.3), ABCi formulation is confirmed to improve ABC's results when the *limit* is lesser than 50, no statistical evidence when limit value increases. Comparing ABCi with k-means we prove better results of honey bee approaches. Moreover, considering Dermatology dataset as depicted in Fig.3, we can observe how *limit* ranging between 20 and 50 represents the best choice even if ABC approach does not outperform k-means due to the colony size which need to be increased as proved in Fig.1.

Finally, in order to study the effect of the number of onlookers for algorithm's convergence speed, we show in Fig.2 the average fitness behavior of 20 different runs. As we expected, the more number of onlookers increases, the more quickly the algorithm converges.

As an example of application in Transportation domain, we consider the problem of vehicle clustering. The purpose of the analysis is to group together Poste Italiane vehicle with the same features, i.e., the average monthly fuel consumption index and the average monthly vehicle route. In particular, fuel consumption index measures the vehicle's cost and identify at the same time the vehicle's performance. Indeed, it considers the fuel demand related to the followed route.

Starting with a set of 10984 Poste Italiane cars which supply postal items to the national addressees, we adopt ABC clustering in order to group together in a same cluster those cars with the same delivery behavior and the same fuel consumption index. The ABC algorithm is setup with following parameters: MCN = 1000, colony size = 61 (50 onlooker bees, 10 employed bees and 1 inspector), *limit* = 50.

Analyzing the results, we can state that the cars are properly grouped in clusters which are suitable for knowledge extraction process and are useful to understand the reason of the provided cars' performances.

7   Conclusions and Future Work

In this paper we presented a bee colony algorithm for clustering problem. Starting from the experiment conducted by Granovskiy regarding the role of inspector bee within a colony, we modeled and proposed a bee colony with inspector. Our experimentation showed the impact in adopting this bee within the colony, and the benefit is proved.

Comparing bee colony with other evolutionary techniques as genetic algorithms, the role of inspector in the convergence can be compared to the role of elitism in genetic approach.

We adopted bee colony for different clustering tasks from biomedical to industrial domain and experimentation provided very encouraging results, proving the ability of a ABC algorithm in converging towards solutions with high fitness, also in presence of different features (e.g., Dermatology dataset) and different input. Moreover, the algorithm has been proven to provide better results increasing the colony size and exploration and exploitation behavior is investigated as long as the *limit* changes.

However, we aim to investigate two main directions in the future. The first is how to improve performance in algorithm's computational time. Parallel ABC colony seems to be a promising solution. The second direction is to investigate other real-world optimization problems with ABC as the vehicle routing problem. In this case, ABC poses additional interesting questions and can be a valid solution in Intelligent Transport System domain.

References

1. Jain, A.K.: Data clustering: 50 years beyond k-means. Pattern Recognition Letters 31(8), 651–666 (2010)
2. Hruschka, E., Campello, R.J.G.B., Freitas, A., De Carvalho, A.C.P.L.F.: A survey of evolutionary algorithms for clustering. IEEE Transactions on Systems, Man, and Cybernetics, Part C: Applications and Reviews 39(2), 133–155 (2009)
3. Karaboga, D., Ozturk, C.: A novel clustering approach: Artificial bee colony (ABC) algorithm. Applied Soft Computing 11(1), 652–657 (2011)
4. Yan, X., Zhu, Y., Zou, W., Wang, L.: A new approach for data clustering using hybrid artificial bee colony algorithm. Neurocomput. 97, 241–250 (2012)
5. Karaboga, D.: An idea based on Honey Bee Swarm for Numerical Optimization. Technical Report TR06, Erciyes University (October 2005)
6. Karaboga, D., Basturk, B.: On the performance of artificial bee colony (ABC) algorithm. Applied Soft Computing 8(1), 687–697 (2008)
7. Abu-Mouti, F., El-Hawary, M.: Overview of artificial bee colony (ABC) algorithm and its applications. In: 2012 IEEE International Systems Conference (SysCon), pp. 1–6 (2012)
8. Biesmeijer, J.C., de Vries, H.: Exploration and exploitation of food sources by social insect colonies: a revision of the scout-recruit concept. Behavioral Ecology and Sociobiology 49(2-3), 89–99 (2001)
9. Granovskiy, B., Latty, T., Duncan, M., Sumpter, D.J.T., Beekman, M.: How dancing honey bees keep track of changes: the role of inspector bees. Behavioral Ecology 23(3), 588–596 (2012)
10. Bache, K., Lichman, M.: UCI machine learning repository (2013)
11. Akay, B., Karaboga, D.: Parameter tuning for the artificial bee colony algorithm. In: Nguyen, N.T., Kowalczyk, R., Chen, S.-M. (eds.) ICCCI 2009. LNCS, vol. 5796, pp. 608–619. Springer, Heidelberg (2009)

Personalized Multi-day Trips to Touristic Regions: A Hybrid GA-VND Approach

Ali Divsalar[1,2], Pieter Vansteenwegen[1], Masoud Chitsaz[1],
Kenneth Sörensen[3], and Dirk Cattrysse[1]

[1] KU Leuven, Center for Industrial Management, Leuven, 3001, Belgium
{Ali.Divsalaer,Pieter.Vansteenwegen,
Dirk.Cattrysse}@cib.kuleuven.be,
Masoud.Chitsaz@student.kuleuven.be
[2] Faculty of Mechanical Engineering,
Babol University of Technology, Babol, Mazandaran, Iran
kenneth.sorensen@ua.ac.be
[3] University of Antwerp, Prinsstraat 13, 2000 Antwerp, Belgium

Abstract. When a tourist is visiting a large region with many attractions, frequently there is not enough time to reach all of them. Moreover when the journey takes more than a day, at the end of each day an accomodation place should be selected to continue the trip the next day. In this research, we introduce the Orienteering Problem with Hotel Selection and Time Windows (OPHS-TW) in order to model this real application. A set of 395 benchmark instances with known optimal solution are created and a hybrid Genetic Algorithm with a Variable Neighborhood Descent (GA-VND) phase is developed to efficiently solve the instances in a reasonable time.

Keywords: Orienteering problem, Hotel selection, Time windows, Genetic algorithm, Variable neighborhood descent.

1   Introduction

Imagine a tourist who wants to visit a large touristic region such as a part of a country or, for instance, Europe in a couple of days. In this case, in each day some of the possible attractions should be selected to visit and also an accommodation place should be chosen to stay each night. Obviously, the location of the accommodation has influence on which possible attractions can be visited during the day. This application of tourist trip planning is modeled as the Orienteering Problem with Hotel Selection (OPHS) in the literature [1][2]. In practice, points of interests (POI) have limited visiting hours. Therefore, we add an opening and closing time for every POI. This leads to a new variant of the OPHS which we call the Orienteering Problem with Hotel Selection and Time Windows (OPHS-TW).

In the OPHS-TW, two sets of nodes are available, H and N. The set of hotels, H, contains h hotels including the initial and the final hotels. The set of POIs contains n

C. Blum and G. Ochoa (Eds.): EvoCOP 2014, LNCS 8600, pp. 194–205, 2014.

nodes, each with a score, S_i. The time needed to travel between each pair of nodes $\in H \cup N$, $t_{i,j}$ is given. The whole multi-day visit plan of the POIs starts from and ends in the given initial and final hotels. This is called a "tour" and is composed of some daily plans each called a "trip". Each trip starts and ends in one of the available hotels and is limited by a maximum available time, T_d. Each POI is assigned a time window, $[O_i, C_i]$. A score is only collected if a POI is visited during its time window. In this case, the arrival at a certain POI$_i$ after C_i will not lead to a score and an early arrival will lead to a waiting time, W_i. Each POI score can be collected at most once. The goal is to maximize the sum of the collected scores in the tour composed of a given number of trips. In our formulation of the OPHS-TW, the time spent at each POI (service time) is not considered. During the modeling of a real problem this time can easily be included in $t_{i,j}$.

Several applications of the OPHS are introduced in the literature[1][2] such as tourist trip planning, a military reconnaissance activity composed of consecutive missions between multiple save zones, truck drivers with limited driving time considering the need to find an appropriate parking space at the end of each day. All these can be modeled more realistically by considering time windows.

In this paper, a hybrid Genetic algorithm with an embedded Variable Neighborhood Descent (GA-VND) algorithm is designed to deal with the two levels of the OPHS-TW: a main level of finding the sequence of hotels to stay at the end of each day/trip, and the sub level of finding a sequence of POIs to be visited in each trip. The GA-VND uses two crossovers and one mutation operator to efficiently find a good sequence of hotels in the tour. The VND part, containing five local search moves, is applied in different steps of the algorithm to find good sequences of POIs between the hotels.

The main contributions of this paper are the introduction of the first method to solve the OPHS-TW as well as the creation of benchmark instances with known optimal solution for this problem. The GA-VND has the same structure of the memetic algorithm developed in [2] to solve the OPHS, but the algorithm is adapted for the version with time windows (OPHS-TW).

The remainder of this paper is organized as follows: A brief literature review is presented in Section 2. The proposed algorithm is discussed in Section 3. Section 4 describes how the instance generation procedure was designed and presents the computational experiments. Section 5 concludes this paper.

2   Related Literature

Although this is the first paper on the OPHS-TW, the OPHS has recently been introduced by Divsalar et al.[1][2]. A Skewed Variable Neighborhood Search (SVNS) method as well as a Memetic Algorithm (MA) which combines a genetic algorithm with a variable neighborhood descent are tested on benchmark instances with known optimal solution. Moreover, the (Team) Orienteering Problem, (T)OP, with time windows ((T)OPTW) is considered in a large number of publications. An iterative three component heuristic for the TOPTW is introduced in [3]. In this method, first a variable neighborhood search (VNS) and a simulated annealing (SA) create a pool of routes. Then, a set packing problem is solved to make a tour over the created routes. In [4] a genetic algorithm is developed to tackle the OPTW. Their genetic algorithm uses a specific mutation based on the idea of insertion and shake introduced by Vansteenwegen et al. [5]. Labadie et al. [6] developed an LP-based granular VNS for the TOPTW. The idea is to increase the efficiency of neighborhood exploring by a pre network analysis called granularity based on the idea of Toth & Vigo [7] on the Vehicle Routing Problem (VRP). An Ant Colony System (ACS) is introduced by Montemanni & Gambardella [8] for the TOPTW. For more literature on the OP and its variants we refer to a recent survey by Vansteenwegen et al. [9]. Other tourist applications modeled by the OP can be found, for instance, in [10] and [11].

Another group of related works to the OPHS(-TW) is mainly called the Vehicle Routing Problems with Intermediate Facilities (VRP-IF) in the literature [12]. An extensive literature survey on this class of problems can be found in [2]. The concept of intermediate facilities is also used in the context of Electric Vehicle Routing Problem (EVRP) when considering the recharging stations as intermediate facilities[13].

3   The Proposed Algorithm

Considering the two-level structure of the OPHS-TW, a two-level algorithm of GA-VND is developed in this paper. In the proposed algorithm, the upper level, genetic algorithm, includes two crossover and one mutation operator, and mainly focuses on creating a good sequence of hotels in the solution. A variable neighborhood descent (VND) is the lower level of the proposed algorithm. This VND is composed of five local search moves designed for finding a good sequence of POIs in each trip taking care of the time window constraints and considering the POIs visited in other trips. To perform these moves efficiently, for every included POI_i in the tour, its arrival time, AT_i, service start time, ST_i and $MaxShift_i$ are stored in memory. The concept of MaxShift (the maximum time that the visit to a POI can be delayed without violating any time window or the trip length) is introduced by Vansteenwegen et al. [5] for the TOPTW. In fact, keeping track of the MaxShift for each POI, makes the feasibility evaluation of an insertion efficient. Also the waiting time, W_i, of each POI, which is calculated by equation (1), is recorded for every visited POI. `

$$W_i = \max(0, O_i - AT_i) \tag{1}$$

3.1   General Structure of the Algorithm

The general structure of the proposed GA-VND is presented in Algorithm 1. The algorithm includes two main phases: the *initialization* and the *main-loop*. In the initialization phase, the initial population is created after performing a preprocessing step. Then, in each iteration of the main-loop, a pool of individuals (Pool) is filled using the GA operators, and the new population is selected from the Pool. The main-loop is stopped only if a maximum number of iterations is reached (*MaxIteration*).

Algorithm 1. General structure of the GA-VND
```
1:    Initialization
2:    Main-loop:
3:              IterN ← 1
4:              While (IterN ≤ MaxIteration) Do
5:              |         Fill the  Pool
6:              |         Sort the Pool according to their fitness
7:              |         Save the "Best − Found − Solution"
8:              |         Select from the Pool:
9:              |         IterN ← IterN + 1
10:             End
11:   Output: "Best − Found − Solution"
```

3.2   Initialization

In this phase, a preprocessing step is done to have an idea about the appropriate hotel selection at the end of each trip. Then, using this information, the initial population is generated.

Matrices of Pairs of Hotels
    For each possible pair of hotels for each trip, a regular orienteering problem is solved applying a very fast "best-insertion" algorithm. The insertion move here is the Insert move explained in Section 3.4. In fact, this move is applied as long as a POI is added to the OP solution. Then, the results are stored in two three-dimensional matrices. Afterwards, for each hotel in each trip, the possible end hotels are sorted based on their score. This sorted list is used in the later steps of the algorithm.

Initial Population
    To create each individual for the initial population, first a feasible sequence of hotels is constructed. A recursive function is used which starts from the initial hotel and randomly selects a hotel from the sorted list of possible end hotels considering their score in the matrix of pairs of hotels. After that, the VND is applied on the sequence of hotels to make a complete OPHS-TW solution. This process is the same as what is done in [2] to create the initial population, but in here the moves are adapted to consider the time windows.

3.3   Genetic Algorithm

After creating the initial population, it is used as the current population and the GA is performed to create a pool of individuals out of it (Pool). The Size of this pool is twice the population size (*PopSize*). In fact, the Pool contains the current population as well as new individuals created by applying the GA operators on the current population. At the end, the new population is selected from the Pool.

Solution Representation
    The GA chromosome representing the OPHS-TW solution is only a list of numbers representing the sequence of hotels in the tour. The fitness of an individual is the total score of the solution, also taking into account the visited POIs.

Genetic Operators

To create new individuals for the Pool, parents are selected from the current population and two crossover and one mutation operator are designed to make offspring out of the parents. Algorithm 2 shows the process of filling the Pool, including the corresponding parameters. These operators are discussed in the following sections. Each parent is selected using a roulette wheel selection method.

Algorithm 2. Fill the Pool

```
1:      Pool ← Current generation
2:      CRI_C ← 1
3:      While (CRI_C ≤ CRI_R × PopSize) Do
4:      |           Select individuals P_1 and P_2 from the Current Generation (Roulette Wheel)
5:      |           Apply Crossover I→ O_1 and O_2
6:      |           Apply Local Search on O_1 and O_2
7:      |           Pool ← O_1 and O_2
8:      |           CRI_C ← CRI_C + 1
9:      End
10:     CRII_C ← 1
11:     While (CRII_C ≤ CRII_R × PopSize) Do
12:     |           Select individuals P'_1 and P'_2 from the Current Generation (Roulette Wheel)
13:     |           Apply Crossover II→ O'_1 and O'_2
14:     |           Apply Local Search on O'_1 and O'_2
15:     |           Pool ← O'_1 and O'_2
16:     |           CRII_C ← CRII_C + 1
17:     End
18:     Mut_C ← 1
19:     While (Mut_C ≤ (PopSize − (CRI_R + CRII_R) × PopSize) Do
20:     |           Select individual P''_1 from the Current Generation (Roulette Wheel)
21:     |           Apply Mutation → O''_1
22:     |           Apply Local Search on O''_1
23:     |           Pool ← O''_1
24:     |           Mut_C ← Mut_C + 1
25:     End
```

Crossovers

There are two crossover operators used in the GA. Both of them are basically creating new sequences of hotels to make two offspring by combining the sequences of hotels in the selected parents. This part in both crossovers is the same and is a one point crossover in which one trip in one parent is randomly selected (pivot trip). If it is feasible to reach from the initial hotel of the pivot trip in the first parent the final hotel of the same trip in the second parent, this combination is performed to have a new sequence of hotels in the offspring. The same process is done for the second parent to be combined with the first using a possibly different pivot trip. In the second part of the crossovers different strategies are applied on the POIs inside the trips.

In Crossover I, a move called "hotel-exchange" is applied on the sequence of hotels in the offspring. This move changes the order of the hotels in the offspring to their best order according to their score in the matrix of pairs of hotels. Then the VND is applied on it to make a complete OPHS-TW solution. Therefore, in Crossover I, basically no information of the visited POIs between the hotels is inherited from the parents.

In Crossover II, the POIs between the hotels in the parents are also used in the offspring. To do that, all the trips until, and including, the pivot trip are copied from the first parent to the offspring. The last hotel of the pivot trip and the remaining trips comes from the second parent. This may cause an infeasibility, in terms of violating the pivot trip length. To repair the trip, the POIs are removed one by one to meet the trip length constraint. The POI with the lowest value of rational score is removed from the pivot trip before the others. The rational score, RS_i for POI_i is calculated using equations (2) and (3). In equation (3), W_i displays the waiting time for POI_i in its current situation, and W'_{i+1} is the waiting time of POI_{i+1} after removing POI_i.

$$RS_i = \frac{(S_i)^2}{delta} \tag{2}$$

$$delta = t_{i-1,i} + W_i + t_{i,i+1} + W_{i+1} - (t_{i-1,i+1} + W'_{i+1}) \tag{3}$$

At the end, the VND is applied on each offspring for further improvements.

Mutation

Each time Mutation is applied on one selected parent from the current population, one or two of its hotels are changed to create a different sequence of hotels. To do this, a trip is randomly selected from the parent (pivot). The final hotel of the pivot trip is updated using the list of possible end hotels of its starting hotel. If using this hotel leads to an infeasible pair of hotels for the trip after the pivot, the ending hotel of that trip is also considered for updating using the same procedure. This process does not go any deeper. To avoid from tapping in a loop of selecting the same ending hotels for a trip, for each trip a number of recently selected hotels are made tabu. This number called *TabuSize*, is a parameter of the algorithm. After updating the hotels, first the POIs in the two/three affected trips are removed and filled with the solution saved for the corresponding trip in the matrix of pairs of hotel. Then, the duplicated POIs in each trip are removed, first from the affected trips and then from the other trips. Removing each POI from a trip needs an update in the arrival time and service start of the following POIs as well as an update in MaxShift of all POIs in the trip.

Selection from the Pool

In this step, the individuals are selected from the Pool to compose the new population. The individuals are selected under three different strategies to keep both high quality and diverse solutions in the population. In the beginning, the individuals in the Pool are sorted based on their fitness. Then, a number of them with the highest fitness are transferred to the new population. This number is equal to $BestSel\_R \times 2 \times PopSize$, in which, $BestSel\_R$ is the percentage of the best individuals in the Pool and is a parameter of the algorithm. *PopSize* shows the population size and is another parameter of the algorithm. After that, from the other individuals of the Pool, only the ones with a different sequence of hotels from the already selected individuals are chosen. If by using this strategy the number of selected individuals for the new population is less than the population size, the remaining individuals are selected from the individuals with the lowest fitness, in order to maintain diversity.

3.4   Variable Neighborhood Descent

The VND part of the algorithm is composed of five local search moves designed to find a good sequence of POIs between the hotels in the tour. Among them, Insert, Extract-Insert and Extract2-Insert are performed to increase the score of the solution. Re-locate and Two-Opt are decreasing the travel time in the tour in order to increase the chance of adding more POIs to the tour. The order of applying the moves and the structure of the VND is presented in Algorithm 3.

Algorithm 3. Local Search
```
1:      X: The incumbent solution of the Local Search
2:      Set of five neighborhood structures (N_k):
3:      Insert, Re-locate, Two-Opt, Extract-Insert, Extract2-Insert;
4:      k ← 0
5:   While (k < 5) Do
6:   |          Apply neighborhood structure N_k on X → X'
7:   |          If X' is better than X then
8:   |          |          X ← X'
9:   |          |          k ← 0
10:  |          Else
11:  |          |          k ← k + 1
12:  |          End
13:  End
```

Insert

   This is a best-improvement move in which for every non-included POI, the position with the minimum increase in trip length ($shift$) is determined and only the POI with the highest ratio of $\frac{s_i^3}{shift_i}$ is inserted in its best position. To evaluate the feasibility of insertionPOI_i between POI_j and POI_{j+1}, $shift_i$ is efficiently calculated using the method introduced in [5]: $Shift_i = t_{j,i} + W_i + t_{i,j+1} - t_{j,j+1}$. Then, the node i can be inserted between nodes j and $j + 1$ if the inequality $Shift_i < W_{j+1} + MaxShift_{j+1}$ is satisfied.

Re-Locate

   For every POI in the trip, it is checked if there is a better position in the whole tour that can lead to less travel time for the considering trip(s). For every possible move of POI_i from its current position (between visiting POI_{i-1} and POI_{i+1}) to a position after POI_k, a feasibility evaluation is performed. This evaluation is not straightforward due to the time windows, but can be performed efficiently thanks to the recording of Wait and MaxShift. When POI_i and POI_k are in the same trip (intra Re-Locate), Algorithm 4 shows the necessary steps for this feasibility evaluation and if the POIs are in different trips (inter Re-Locate) the calculations are shown by Algorithm 5. In this neighborhood search, only the POI with the highest value for $shift_i$ is moved. $Shift_i$ is calculated using equation (4).

$$Shift_i = t_{i-1,i} + t_{i,i+1} - t_{i-1,i+1} + t_{k,k+1} - t_{k,i} - t_{i,k+1} \qquad (4)$$

Algorithm 4. Feasibility evaluation in Intra Re-Locate

```
1:    AT'ᵢ₊₁ ← STᵢ₋₁ + tᵢ₋₁,ᵢ₊₁
2:    W'ᵢ₊₁ ← max (0, Oᵢ₊₁ − AT'ᵢ₊₁)
3:    ST'ᵢ₊₁ ← AT'ᵢ₊₁ + W'ᵢ₊₁
4:    For (j = i+2 to j = k)
5:    |        AT'ⱼ ← ST'ⱼ₋₁ + tⱼ₋₁,ⱼ
6:    |        W'ⱼ ← max (0, Oⱼ − AT'ⱼ)
7:    |        ST'ⱼ ← AT'ⱼ + W'ⱼ
8:    |        If (ST'ⱼ = STⱼ) Break
9:    End
10:   AT'ᵢ ← ST'ₖ + tₖ,ᵢ
11:   If (AT'ᵢ < Cᵢ)
12:   |        W'ᵢ ← max (0, Oᵢ − AT'ᵢ)
13:   |        ST'ᵢ ← AT'ᵢ + W'ᵢ
14:   |        AT'ₖ₊₁ ← ST'ᵢ + tᵢ,ₖ₊₁
15:   |        W'ₖ₊₁ ← max (0, Oₖ₊₁ − AT'ₖ₊₁)
16:   |        ST'ₖ₊₁ ← AT'ₖ₊₁ + W'ₖ₊₁
17:   |        If (ST'ₖ₊₁ − STₖ₊₁ < MaxShiftₖ₊₁)
18:   |        |        The move is feasible
19:   |        |        Save the necessary information
20:   |        End
21:   End
```

Two-Opt

To apply this move, the first-improvement strategy is used. Starting from the first trip, for each pair of visited POIs, if an inversion in the visit order of POIs between them leads to a saving in travel time and the move is feasible, then it is performed. Due to the time windows, the feasibility evaluation is done in $O(n)$ in its worst case.

Extract-Insert

For every POI in every trip, starting from the first visited POI in the first trip, if by excluding this POI and inserting as many POIs as possible in the same trip, it is possible to increase the total score, the move is performed and the next POI is considered.

To do the feasibility evaluation, after removing a POI from a trip, all the modified time window characteristics of all the POIs in the same trip are calculated once. Then, *Insert* is evaluated and applied as many times as possible using the same algorithm explained above.

Extract2-Insert

This move is similar to Extract-Insert considering that each time two consecutive POIs are selected for exclusion.

Algorithm 5. Feasibility evaluation in Inter Re-Locate

```
1:    AT'ᵢ ← ST'ₖ + tₖ,ᵢ
2:    If (AT'ᵢ < Cᵢ)
3:    |        W'ᵢ ← max (0, Oᵢ − AT'ᵢ)
4:    |        ST'ᵢ ← AT'ᵢ + W'ᵢ
5:    |        AT'ₖ₊₁ ← ST'ᵢ + tᵢ,ₖ₊₁
6:    |        W'ₖ₊₁ ← max (0, Oₖ₊₁ − AT'ₖ₊₁)
7:    |        ST'ₖ₊₁ ← AT'ₖ₊₁ + W'ₖ₊₁
8:    |        If (ST'ₖ₊₁ − STₖ₊₁ < MaxShiftₖ₊₁)
9:    |        |        The move is feasible
10:   |        |        Save the necessary information
11:   |        End
12:   End
```

4   Computational Experiments

First we explain how the benchmark instances are created and then the computational results are presented and discussed in detail.

4.1   Benchmark Instances

In [1][2] it is explained how benchmark instances of OPHS with known optimal solutions are created based on optimal solutions of OP instances. Adding time windows to these benchmark instances, a set of 395 OPHS-TW instances are created. To create these new OPHS-TW instances from the OPHS benchmark instances, the set of POIs in the OPHS, N, are divided into two subsets: POIs which are visited in the OPHS optimal solution, N_{opt}, and the ones not included in the optimal solution, N_{nop}. For each $POI_i \in N_{opt}$, the arrival time, AT_i, in the optimal solution is determined. .Then, the O_i and C_i are calculated using equations (5) and (6) in which r_1 and r'_1 are two uniformly generated random numbers in $[0,1]$.

$$O_i = \begin{cases} 0; \ if \ AT_i - (0.05 + 0.2 \times r_1) \times T_d < 1 \\ AT_i - (0.05 + 0.2 \times r_1) \times T_d; Otherwise \end{cases} \tag{5}$$

$$C_i = \begin{cases} 1; \ if \ AT_i + (0.05 + 0.2 \times r'_1) \times T_d < 1 \\ AT_i + (0.05 + 0.2 \times r'_1) \times T_d; Otherwise \end{cases} \tag{6}$$

To create random time window for a $POI_i \in N_{nop}$, first T_{avg} and AT'_i are calculated using the equation (7) and (8). Then, equations (9) and (10) are used to calculate the O_i and C_i for these POIs.

$$T_{avg} = \frac{\Sigma_{i=1}^{D} T_i}{D} \tag{7}$$

$$AT'_i = r_2 \times T_{avg} \tag{8}$$

$$O_i = \begin{cases} 0; \ if \ AT'_i - (0.05 + 0.2 \times r_3) \times T_{avg} < 1 \\ AT'_i - (0.05 + 0.2 \times r_3) \times T_{davg}; Otherwise \end{cases} \tag{9}$$

$$C_i = \begin{cases} 1; \ if \ AT'_i + (0.05 + 0.2 \times r'_3) \times T_{avg} < 1 \\ AT'_i + (0.05 + 0.2 \times r'_3) \times T_{avg}; Otherwise \end{cases} \tag{10}$$

r_2, r_3 and r'_3 are also uniformly generated random numbers in $[0,1]$.

In this way, the optimal solution of the OPHS instance remains optimal also for the OPHS-TW instance. This results from the fact that adding extra constraints (time windows in this case) to an optimization problem can never result in a better solution. Moreover, we added the time windows to an OPHS instance in such a way that its optimal solution is still feasible for the created OPHS-TW instance. Therefore, it is not possible for the OPHS-TW to find a better solution than the optimal solution of the OPHS. The instances and the detailed results of applying the GA-VND are available at http://www.mech.kuleuven.be/en/cib/op.

4.2   Results

All computations are carried out on a Intel Core 2 desktop with 3.00 GHz processor and 4.00 GB RAM. The parameters of our algorithm are tuned using a number of preliminary experiments and set to *MaxIteration*= 100, *PopSize*= 30, *CRI_R*= 30%, *CRII_R*= 40% , *BestSel_R*= 25%   and *TabuSize*=33% of *h*.

The benchmark instances are categorized into 16 sets based on the number of hotels as well as the number of trips in the tour. Due to randomness in the algorithm, the GA-VND is applied three times to each instance and its best and average results are recorded. Then, to present the results in Table 1, for each set, the average gap and the maximal gap for both the average and the best results out of three runs are presented in columns 5 to 8. Columns 1 to 4 in Table 1 present the set number, the number of "intermediate" hotels (excluding the initial and final hotel), the number of trips in each instance and the total number of instances in each set.

Table 1. The results of the experiments

Set Name			Number of instances	Average Gap		Maximum Gap		Avg CPU
Set number	Intermediate hotels	Trips		Best	Avg	Best	Avg	
1	1	2	35	1.83	2.33	27.76	27.76	0.77
2	2	3	35	0.46	0.72	4.88	7.08	0.54
3	5	3	35	0.51	0.77	4.11	7.72	0.53
4	3	4	35	0.66	0.75	7.51	7.51	0.43
5	6	4	35	0.44	0.61	4.67	4.67	0.43
6	10	4	22	0.72	0.92	3.65	4.27	2.03
7	10	5	22	0.78	1.11	4.61	4.61	1.60
8	10	6	22	0.82	1.37	4.79	7.69	1.31
9	12	4	22	0.47	0.75	3.16	3.16	2.01
10	12	5	22	0.59	1.49	4.61	6.25	1.64
11	12	6	22	0.89	1.81	3.90	6.74	1.33
12	15	4	22	0.53	1.22	3.16	4.18	2.04
13	15	5	22	0.59	1.77	4.61	5.88	1.58
14	15	6	22	0.73	1.69	3.90	6.26	1.34
15	15	8	13	3.48	5.99	8.88	9.98	1.33
16	15	10	9	2.58	6.23	9.08	12.96	1.32
Total			395	1.00	1.84	27.76	27.76	1.26

Since the presented algorithm is the first to deal with the OPHS-TW, the results are compared with the optimal solution.

The average Best gap over all the instances considering the best out of three runs, is only 1.00 % and the average computational time (per run) is 1.26 seconds which is both reasonable in case of the tourist application. Looking at the results, it is evident that the worst gap happens when the number of possible sequences of hotels is very low. This can be explained by the structure of the GA-VND. A lot of effort goes to finding the best

sequence of hotels, which is rather useless for these small instances. Only limited efforts
go to solving the combination of OPs which is very important for these small instances.
On the other hand, generally, when the number of intermediate hotels as well as the number of trips are increased (for instance in set 15 and 16), finding a good sequence of hotels
gets more difficult and therefore the OPHS-TW instances become more difficult to solve.

5   Conclusion

In this paper, the OPHS-TW is introduced. A Set of 395 instances with known optimal solutions are created and a hybrid genetic algorithm (GA-VND) is developed.
The computational results show that the algorithm is able to solve the OPHS-TW
instances effectively in an acceptable time for a tourist application. The two-level
structure of the algorithm perfectly fits to the two-level structure of the problem.
Moreover, integrating diversification in designing crossovers and mutation operators
makes the algorithm able to solve the combination of OPs between the hotels effectively. In designing the neighborhood moves, the concept of MaxShift and Waiting
time is used to make the algorithm computationally efficient.

For the future work, another framework should be developed in order to better
solve the larger instances and to solve other variants of OPHS-TW, for instance, when
the hotels and/or arcs also have a score. Furthermore, although we can evaluate the
performance of our algorithm by comparing our results with available optimal
solutions, implementing another solution framework could give extra insight in the
difficulty of the OPHS-TW instances we created.

References

1. Divsalar, A., Vansteenwegen, P., Cattrysse, D.: A Variable Neighborhood Search Method
   for the Orienteering Problem with Hotel Selection. Int. J. Prod. Econ. 145, 150–160 (2013)
2. Divsalar, A., Vansteenwegen, P., Sörensen, K., Cattrysse, D.: A Memetic Algorithm for
   the Orienteering Problem with Hotel Selection. Eur. J. Oper. Res. (2014)
3. Hu, Q., Lim, A.: An iterative three-component heuristic for the team orienteering problem
   with time windows. Eur. J. Oper. Res. 232, 276–286 (2014)
4. Koszelew, J., Ostrowski, K.: A Genetic Algorithm with Multiple Mutation which Solves
   Orienteering Problem in Large Networks. In: Bădică, C., Nguyen, N.T., Brezovan, M.
   (eds.) ICCCI 2013. LNCS, vol. 8083, pp. 356–366. Springer, Heidelberg (2013)
5. Vansteenwegen, P., Souffriau, W., Vanden Berghe, G., Van Oudheusden, D.: Iterated local
   search for the team orienteering problem with time windows. Comput. Oper. Res. 36,
   3281–3290 (2009)
6. Labadie, N., Mansini, R., Melechovsky, J., Wolfler Calvo, R., Melechovský, J.: The Team
   Orienteering Problem with Time Windows: An LP-based Granular Variable Neighborhood
   Search. Eur. J. Oper. Res. 220, 15–27 (2012)
7. Toth, P., Vigo, D.: The Granular Tabu Search and Its Application to the Vehicle-Routing
   Problem. INFORMS J. Comput. 15, 333–346 (2003)
8. Montemanni, R., Gambardella, L.M.: AN Ant Colony System for Team Orienteering Problems with Time Windows. Found. Comput. Decis. Sci. 34, 287–306 (2009)

9. Vansteenwegen, P., Souffriau, W., Van Oudheusden, D.: The orienteering problem: A survey. Eur. J. Oper. Res. 209, 1–11 (2011)
10. Schilde, M., Doerner, K.F., Hartl, R.F., Kiechle, G.: Metaheuristics for the bi-objective orienteering problem. Swarm Intell. 3, 179–201 (2009)
11. Vansteenwegen, P., Souffriau, W., Vanden Berghe, G., Van Oudheusden, D.: The City Trip Planner: An expert system for tourists. Expert Syst. Appl. 38, 6540–6546 (2011)
12. Angelelli, E., Speranza, M.G.: The periodic vehicle routing problem with intermediate facilities. Eur. J. Oper. Res. 137, 233–247 (2002)
13. Schneider, M., Stenger, A., Goeke, D.: The Electric Vehicle Routing Problem with Time Windows and Recharging Stations. Kaiserslautern (2012)

Phase Transition and Landscape Properties of the Number Partitioning Problem

Khulood Alyahya and Jonathan E. Rowe

School of Computer Science
University of Birmingham, B15 2TT, UK

Abstract. This paper empirically studies basic properties of the fitness landscape of random instances of number partitioning problem, with a focus on how these properties change with the phase transition. The properties include number of local and global optima, number of plateaus, basin size and its correlation with fitness. The only two properties that were found to change when the problem crosses the phase transition are the number of global optima and the number of plateaus, the rest of the properties remained oblivious to the phase transition. This paper, also, studies the effect of different distributions of the weights and different neighbourhood operators on the problem landscape.

Keywords: combinatorial optimisation, phase transition, partitioning problem, makespan scheduling, fitness landscape.

1   Introduction

Many NP-hard problems, like satisfiability (SAT) and graph-colouring problems, have been shown to have a phase transition, which is an analogy to the transitions observed in thermodynamic systems. A phase transition phenomenon in a physical system can be described as an abrupt change of its macroscopic properties at a certain value of the control parameter [6]. A simple example is the transition from ice to water at a specific temperature value. Similar behaviour has been observed in combinatorial problems where a small change in the control parameter introduces a change in the problem properties. In graph-colouring problem, for instance, a small change in the average node degree causes changes in the graph connectivity.

There has been much interest in using the phase transition as a paradigm to study and describe problem difficulty [6]. The most notable example of a well-studied phase transition is the transition in SAT problem, which has an easy-hard-easy phase transition. Problem instances below the threshold where there are few constraints are almost always soluble, while problem instances above the threshold where there are too many constraints are almost always insoluble. The most difficult problem instances lie in the threshold of the phase transition between under-constrained and over-constrained. For Max-Sat, the optimisation version of the satisfiability problem, the instances get harder at the threshold and stay hard above the threshold [6].

C. Blum and G. Ochoa (Eds.): EvoCOP 2014, LNCS 8600, pp. 206–217, 2014.

Such changes in problem difficulty are often accompanied with changes in the structure of the landscape which demonstrates that instances belonging to the same class could have very different landscape structures. In k-SAT, for example, it has been shown that in the hard phase of the problem, the local optima are grouped into different clusters which are far away from each other [10]. It also has been shown that around the phase transition of Max-Sat, there exist large regions of plateau which were found to significantly slow down local search algorithms [14].

A similar phase transition has been identified in the number partitioning problem. In the literature, the effect of this phase transition has been shown in the computational complexity of exact solvers such as the complete Karmarkar-Karp differencing algorithm [9]. In this paper we are interested to see whether the landscape properties of number partitioning problem change with the phase transition. We are also interested to see if similar changes in the computational complexity occur in the performance of local search algorithms. The effect of different distributions of the weights and different neighbourhood operators on the problem landscape were also studied.

2   Number Partitioning Problem

2.1   Problem Definition

The number partitioning problem (NPP) is defined as follows: given a set $A = \{a_1, \ldots, a_n\}$ of positive integers drawn at random from the set $\{1, 2, .., M\}$, the goal is to partition A into two disjoint subsets S, S' such that the discrepancy between them $|\sum_{a_i \in S} a_i - \sum_{a_i \in S'} a_i|$ is minimised. A partition is called perfect, if the discrepancy between the two subsets is, 0 when the sum of the original set is even, or 1 when the sum is odd. Equivalently, the problem can be viewed as minimising the maximum sum over the two subsets.

Let $x \in \{0, 1\}^n$, the fitness function to be minimised can be defined as:

$$f(x) = max \left\{ \sum_{a_i \in A} a_i x_i, \sum_{a_i \in A} a_i (1 - x_i) \right\} \tag{1}$$

Partitioning is a classical problem in theoretical computer science and it is one of Garey and Johnson's six basic NP-complete problems [4]. Throughout this paper we interchangeably use numbers and weights to refer to the integers in the set.

2.2   Phase Transition in NPP

NPP undergoes a sudden phase transition from "hard" to "easy" regimes, determined by the control parameter $k = log_2 M/n$, which corresponds to the number of the bits required to encode the numbers in the set divided by the size of the set. For $log_2 M$ and n tending to infinity, the transition occurs at the critical value of $k_c = 1$, such that for $k < 1$, there are many perfect partitions with probability tending to 1, whereas for $k > 1$, the number of perfect partitions drops to zero with probability tending to 1 [1]. A more detailed parameterisation of the critical value of the control parameter is given by the following[1] [9]:

$$k_c = 1 - \frac{\ln(\frac{\pi}{6}n)}{2n \ln 2} \tag{2}$$

The problem experiences an easy regime when $k < k_c$ and a hard regime when $k > k_c$.

Gent and Walsh were the first to verify the existence of a phase transition in NPP and they have shown an empirical evidence of it in their paper through numerical simulations. They introduced the control parameter k and estimated the transition point to occur around $k_c = 0.96$ [5]. Previously, Fu [3] used statistical mechanics to analyse the problem and concluded incorrectly that NPP does not undergo a phase transition. Mertens [9,8] used the same method from statistical physics and the parameterisation of Gent and Walsh to obtain non-rigorous analytical results of the phase transition in NPP. Borgs et al. [1] then performed a rigorous analysis of the problem and showed the mathematical proofs for the existence of phase transition in NPP.

3   Landscape of NPP

We are interested in finding what makes NPP difficult in the hard phase for local search algorithms. In particular, we want to check if some of the properties of the fitness landscape that have been conjectured to be related to the problem difficulty such as the number of local optima, the size of the global and local basins and the correlation between the basin size and fitness [11,7], change as we go from the easy phase to the hard phase. We are also interested in whether the distribution of the weights has any effect on the problem landscape. Two different neighbourhood operators will be investigated as well, to get an intuition of what could be a more suitable neighbourhood structure.

The first part of this section gives definitions of some of the concepts used in this study and describes the experimental settings. The results of experiments are presented and discussed in the second part.

3.1   Definitions and Experimental Setup

A fitness landscape [13] of a combinatorial optimisation problem is a triple (X, N, f), where X is the finite set of all the admissible solutions, f is the objective function $f : X \to R$, and N is the neighbourhood operator function $N : X \to P(X)$. The neighbourhood operator defines how the solutions are connected in the landscape, and how one can move from one solution to another.

---

[1] A more rigorous derivation of the transition point can be found in [1].

Optima and Plateaus. A point $x \in X$ is a local optimum iff $\forall y \in N(x), f(y) \geq f(x)$. A point $x \in X$ is a strict local optimum iff $\forall y \in N(x), f(y) > f(x)$. A local optimum x is a global optimum iff $\forall y \in X, f(x) \leq f(y)$. A group of connected local optima form a closed plateau [2].

Basin of Attractions. For an optimum $x^* \in X$, its basin of attraction $B(x^*)$ is the set of points that leads to it after applying local search algorithm to them, $B(x^*) = \{x \in X | localsearch(x) = x^*\}$. The local search algorithm to determine the basin of attraction is given in Algorithm 1. If there is more than one neighbour with the best improving move, the first one is always selected.

Algorithm 1. Steepest Descent

---

for $x \in X$
repeat
   choose $x' \in N(x)$, *such that* $f(x') = min_{y \in N(x)} f(y)$
   replace x *with* x' *if* $f(x') < f(x)$
until
x is a strict local optimum or a local optimum

In the experiments, instances from small problem size were considered to allow exhaustive enumeration of the entire search space, the problem size considered is $n = 20$. Since the studied problem is a pseudo-Boolean function the search space size is 2^n. The binary representation of NPP creates a symmetry in the search space, in the sense that a solution and its bitwise complement have the same fitness value. The number of unique solutions is thus 2^{n-1}.

In this paper, we considered two different neighbourhood operators. The 1 hamming operator (H1), the neighbourhood using this operator is the set of points that are reached by 1-bit flip mutation of the current solution, hence the neighbourhood size is $|N(x)| = n$. The second operator is the 1+2 hamming operator (H1+2), the neighbourhood here includes the hamming one neighbours plus the hamming two neighbours of the current solution which can be reached by 2-bits flip mutation, the neighbourhood size for this operator is $|N(x)| = n + (n(n-1)/2)$. 30 instances were generated randomly for each combination of the following experimental settings:

Phase Transition. To study the effect of phase transition on the properties of NPP landscape, the control parameter k was varied between 0.4 to 1.3 with 0.1 step interval.

Different Distributions. Here we are interested in how the distributions of the weights might affect the structure of the problem landscape. Problem instances were generated with weights drawn randomly from five different discrete probability distributions: uniform, normal, negatively skewed, positively skewed and bimodal distribution with peaks at both ends (figure 1 shows example distributions).

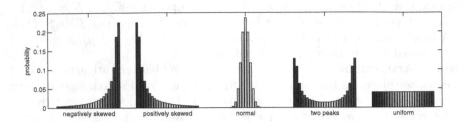

Fig. 1. Illustration of the probability mass functions (pmfs) of the considered distributions of the weights

3.2 Experimental Results

Number of Global Optima and Plateaus. Figure 2 shows the number of global optima for the randomly generated instances of NPP. The figure shows that for all the distributions, the number of global optima decreases as we approach the phase transition point and keep decreasing as we cross the phase transition to only two. There are some variations in the number of global optima between the different distributions in the easy phase, with instances drawn from the positively skewed distribution having the highest number of global optima and instances drawn from the negatively skewed distribution having the lowest number of global optima. Similar results have been observed for the number of points in closed plateaus as figure 3 shows. The figure shows that the number starts to decrease as we approach the phase transition until it becomes zero in the hard phase.

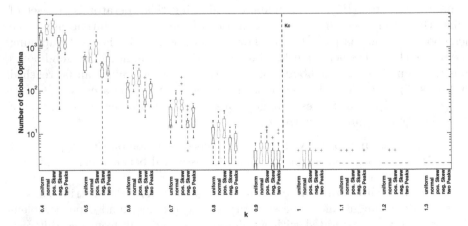

Fig. 2. Number of global optima versus the phase transition control parameter k, for all the different distributions of the weights. Each box represents the number of global optima found in 30 random instances of size $n = 20$. The dotted line is given by k_c from Eq. (2).

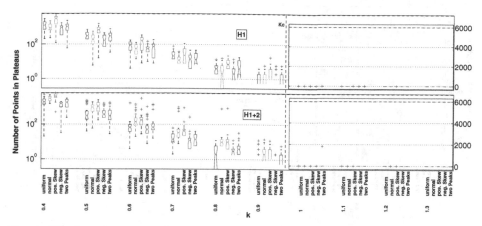

Fig. 3. Number of points in closed plateaus versus the phase transition control parameter k, for all the different distributions of the weights. Each box represents the number of points in plateaus found in the 30 random instances of size $n = 20$. This is shown for both neighbourhood operators H1 and H1+2. The dotted line is given by k_c from Eq. (2).

Number of Local Optima. Figure 4 shows the number of strict local optima (excluding global optima) found in the randomly generated instances of NPP. There is a very clear difference in the number of local optima across instances generated from the different distributions. In the landscapes induced by the H1 operator, instances drawn from normal distribution have the highest number of local optima (around 15% of the search space). Instances generated from negatively skewed distribution have a quite high number of local optima as well (around 8% of the search space) but the number varies a lot between the randomly drawn instances from this distribution. Instances drawn from the uniform distribution have less number of local optima (around 3% of the search space), while the lowest number of local optima is seen in instances drawn from both positively skewed and two peaks distributions (representing around 1% of the search space). Figure 4, also, shows that the number of local optima does not change very much between the easy and the hard phase regardless of the distribution from which the weights are chosen.

Number of local optima can be used as a measure of the ruggedness of the landscape [13]. Thus, the results found indicate that instances with weights drawn from normal and negatively skewed distributions have more rugged landscapes than instance drawn form uniform, positively skewed and two peaks distributions which seem to have much less rugged landscapes. Other features such as the distribution of local optima could be studied in the future to support this observation.

For the landscapes induced by H1+2 operator, the number of local optima drops for all the different distributions compared to their H1 landscapes. It seems that the largest drop occurs in instances drawn from normal and negatively

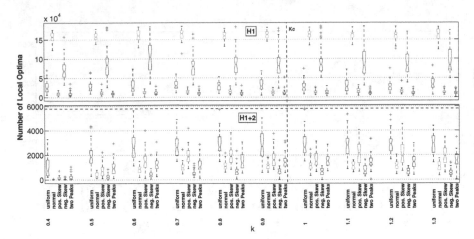

Fig. 4. Number of strict local optima versus the phase transition control parameter k, for all the different distributions of the weights. Each box represents the number of local optima found in the 30 random instances of size $n = 20$. This is shown for both neighbourhood operators H1 and H1+2. The dotted line is given by k_c from Eq. (2).

skewed distributions. As in the H1 landscapes, the number of local optima does not seem to change much between the easy and the hard phase except for very small values of k (0.4 and 0.5) and this could be due to the slightly higher number of points that are part of closed plateaus in the H1+2 landscapes of those instances.

Basin Size and Its Correlation with Fitness. Here we look at another feature of the fitness landscape of NPP, that is, the basins of attraction. Although the number of local optima was found to be insensitive to the phase transition, we are interested here to see if the size of the local optima basins becomes larger in the hard phase resulting in the size of the global basins to shrink, which could be one of the reasons for the problem to be difficult in the hard phase. Figures 5 and 6 show the average basin sizes of the global and local optima respectively. In the H1+2 landscape, there is a large increase in the average basin size in comparison to the H1 landscape. This increase is in accordance with the decrease in the number of local optima for all the different distributions and for all the different values of k, in H1+2 landscape compared to H1 landscape. The figures, also, show that there is not much difference between the average basin sizes in the easy phase and the hard phase for all the different distributions and for both H1 and H1+2 landscapes, given evidence that even the size of the basins seems to be insensitive to the phase transition.

Another important aspect of the fitness landscape is the correlation between the basin size and the fitness of the optimum. Previous studies have shown that in general, fitter optima have larger basins [13]. Here we want to examine if this is also the case in NPP. Figure 7 shows that for the H1 landscape, all instances with weights generated from all the distributions, have negative correlations

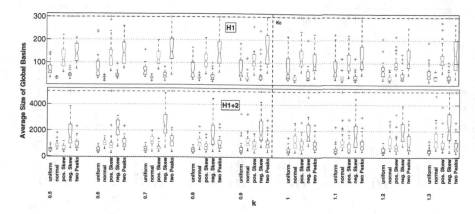

Fig. 5. Average basin size of global optima versus the phase transition control parameter k, for all the different distributions of the weights. Each box represents data from the 30 random instances of size $n = 20$. This is shown for both neighbourhood operators H1 and H1+2. The dotted line is given by k_c from Eq. (2).

between the basin size and fitness, indicating that fitter optima tend to have bigger basins. This, again, does not seem to change very much across the different values of k. For the H1+2 landscape, the negative correlation seems to remain the same for all the distributions except for instances generated from normal and positively skewed distributions. In the normal distributions the correlation fluctuates between positive and negative values, while for instances generated from the positively skewed distribution the correlation occasionally becomes positive.

Cost of Local Search. To examine how the cost of finding the optimal solutions varies from the easy phase to the hard phase, steepest descent with random restart algorithm (Algorithm 2 below) was run with the two neighbourhood operators for 100 times for each instance. The cost of finding the global optima is then calculated using the number of used fitness evaluations.

Algorithm 2. Steepest Descent with Random Restarts

---
repeat
    Chose $x \in \{0,1\}^n$, uniformly at random
    repeat
        choose $x' \in N(x)$, such that $f(x') = min_{y \in N(x)} f(y)$
        replace x with x' if $f(x') < f(x)$
    until
    x is a strict local optimum or a local optimum
until
x is a global optimum

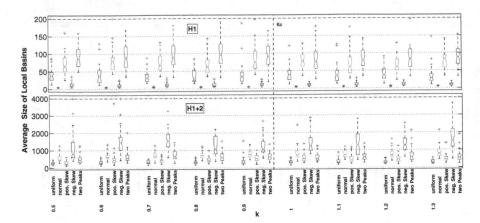

Fig. 6. Average basin size of local optima versus the phase transition control parameter k, for all the different distributions of the weights. Each box represents data from the 30 random instances of size $n = 20$. This is shown for both neighbourhood operators H1 and H1+2. The dotted line is given by k_c from Eq. (2).

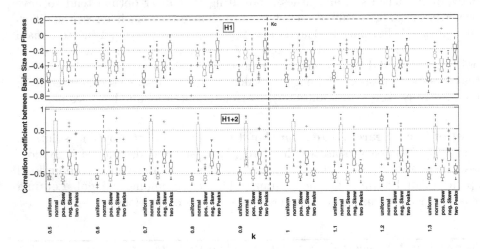

Fig. 7. Correlation coefficient between fitness and basin size versus the phase transition control parameter k, for all the different distributions of the weights. Each box represents data from the 30 random instances of size $n = 20$. This is shown for both neighbourhood operators H1 and H1+2. The dotted line is given by k_c from Eq. (2).

Figure 8 shows the results of the algorithm runs with the H1 and the H1+2 operators. The results were averaged over the 30 problem instances. The Student's t-test was used to determine if the difference between the performance of the H1 and the H1+2 operators is significantly different. In most of the cases, the difference between the performance of the two operators was found to be statistically significant ($p < 0.05$).

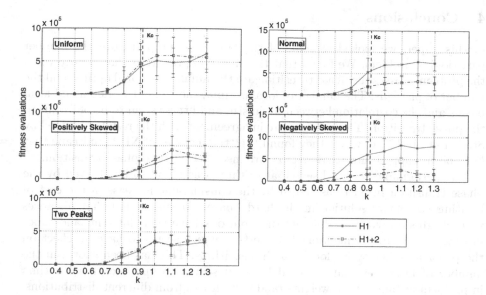

Fig. 8. Average number of fitness evaluations used to find the global optimum, plotted against the phase transition control parameter k. This is shown for all the considered distributions of the weights and for both neighbourhood operators H1 and H1+2. Each data point represents the average over the 30 instances of size $n = 20$ and 100 runs of the steepest descent algorithm per instance. The dotted line is given by k_c from Eq. (2).

For all the different distributions the figure shows that the average number of fitness evaluations used to find the global optima increases as we approach the phase transition point and keep increasing as we cross the phase transition. This is expected due to the drastic decrease in the number of global optima in the hard phase. In the easy phase, the algorithm usually finds quickly one of the many global optima while it struggles to find the single (two if we considered the symmetry) global optimum in the hard phase.

The number of used fitness evaluations varies across the different distributions, instances drawn form positively skewed and two peaks distributions have the lowest number of fitness evaluations, which is unsurprising due to the low number of local optima in the landscape of both cases. For instances drawn from normal and negatively skewed distributions the performance of the H1+2 operator was much better than the performance of the H1 operator. This can be explained by the very big difference between the number of local optima in the H1+2 landscape compared to the H1 landscape which has far more local optima, suggesting that the algorithm probably had to do far less restarts when using the H1+2 operator. For the rest of the distributions, the H1 operator seems to have a better performance even though the number of local optima is less in the landscapes induced by the H1+2 operator. This perhaps can be explained by the number of fitness evaluations needed to explore the much larger neighbourhood of the H1+2 operator, which might have offset the advantage of having less number of local optima.

4 Conclusions

In this paper, we looked at some of the landscape properties of the number partitioning problem. We found that the only two properties that change when the problem crosses the phase transition are the number of global optima and the number of plateaus. The number of local optima, the size of the global and local basins and the correlation between basin size and fitness seem to be oblivious to the phase transition. These results are in agreement with the results obtained by Stadler et al. [12], in which they found that the features of NPP landscape that have been mapped into barriers trees are insensitive to the phase transition.

The performance of local search algorithms was found to be affected by the phase transition in NPP, as shown by the considerable increase in the cost of locating the global solution in the hard phase. We also found that instances with weights drawn from different distributions have very different landscapes in terms of the above mentioned properties, which seems to have an effect on the problem difficulty for local search algorithms. This has been shown in the number of fitness evaluations used by local search to find the global optima in problem instances with weights randomly drawn from different distributions. Two neighbourhood operators were investigated to get an intuition of what could be a more suitable neighbourhood structure for a given problem instance. The results show that local search algorithm with H1 neighbourhood operator preforms better than with H1+2 operator for instances drawn form uniform, positively skewed and two peaks distributions, while the H1+2 operator preformed better for instances with weights drawn from negatively skewed and normal distributions. The results could be used for devising suitable and effective neighbourhood operators depending on the distribution of the weights in a given problem instance.

References

1. Borgs, C., Chayes, J., Pittel, B.: Phase transition and finite-size scaling for the integer partitioning problem. Random Structures & Algorithms 19(3-4), 247–288 (2001)
2. Frank, J., Cheeseman, P., Stutz, J.: When gravity fails: local search topology. Journal of Artificial Intelligence Research 7, 249–281 (1997)
3. Fu, Y.: The use and abuse of statistical mechanics in computational complexity. In: Stein, D.L. (ed.) Lectures in the Sciences of Complexity, vol. 1, pp. 815–826. Addison-Wesley, Reading (1989)
4. Garey, M.R., Johnson, D.S.: Computers and Intractability: A Guide to the Theory of NP-Completeness. Series of books in the mathematical sciences. W.H. Freeman (1979)
5. Gent, I.P., Walsh, T.: Analysis of heuristics for number partitioning. Computational Intelligence 14(3), 430–451 (1998)
6. Hartmann, A.K., Weigt, M.: Phase Transitions in Combinatorial Optimization Problems. John Wiley & Sons (2006)
7. Kallel, L., Naudts, B., Reeves, C.R.: Properties of fitness functions and search landscapes. Theoretical Aspects of Evolutionary Computing, 175–206 (2001)

8. Mertens, S.: Phase transition in the number partitioning problem. Physical Review Letters 81(20), 4281–4284 (1998)
9. Mertens, S.: A physicist's approach to number partitioning. Theoretical Computer Science 265(1-2), 79–108 (2001)
10. Mézard, M., Mora, T., Zecchina, R.: Clustering of solutions in the random satisfiability problem. Physical Review Letters 94, 197205 (2005)
11. Pitzer, E., Affenzeller, M.: A comprehensive survey on fitness landscape analysis. In: Fodor, J., Klempous, R., Suárez Araujo, C.P. (eds.) Recent Advances in Intelligent Engineering Systems. SCI, vol. 378, pp. 161–191. Springer, Heidelberg (2012)
12. Stadler, P.F., Hordijk, W., Fontanari, J.F.: Phase transition and landscape statistics of the number partitioning problem. Physical Review E 67(5), 056701 (2003)
13. Stadler, P.F., Stephens, C.R.: Landscapes and effective fitness. Comments on Theoretical Biology 8(4-5), 389–431 (2002)
14. Tayarani, M., Prugel-Bennett, A.: On the landscape of combinatorial optimisation problems. IEEE Transactions on Evolutionary Computation PP(99) (2013)

The Firefighter Problem: Application of Hybrid Ant Colony Optimization Algorithms*

Christian Blum[1,2], Maria J. Blesa[3], Carlos García-Martínez[4],
Francisco J. Rodríguez[5], and Manuel Lozano[5]

[1] Dept. of Computer Science and Artifical Intelligence, Univ. of the Basque Country
UPV/EHU, San Sebastian, Spain
christian.blum@ehu.es
[2] IKERBASQUE, Basque Foundation for Science, Bilbao, Spain
[3] ALBCOM Research Group, Univ. Politécnica de Catalunya, Barcelona, Spain
mjblesa@lsi.upc.edu
[4] Dept. of Computing and Numerical Analysis, Univ. of Córdoba, Spain
cgarcia@uco.es
[5] Dept. of Computer Science and Artificial Intelligence, Univ. of Granada, Spain
fjrodriguez@decsai.ugr.es, lozano@decsai.ugr.es

Abstract. The firefigther problem is a deterministic discrete-time model
for the spread (and the containment) of fire on an undirected graph. Assuming that the fire breaks out at a predefined set of vertices, the goal is
to save as many vertices as possible from burning. The same model has
also been used in the literature for the simulation of the spreading of deseases. In this work we present, to our knowledge, the first metaheuristics
for tackling this problem. In particular, a pure ant colony optimization
approach and a hybrid variant of this algorithm are proposed. The results show that the hybrid ant colony optimization variant is superior
to the pure ant colony optimization version and to a mathematical programming solver, especially when the graph size and density grows.

1 Introduction

The firefighter problem was initially proposed in 1995 by Hartnell [21] as a
deterministic discrete-time model for the spread (and containment) of fire. Since
then it has also been used, for example, to model the spreading of deseases and
the containment of floods. The firefighter problem has been subject to a wide
variety of research during the last 10-20 years. However, most of this research was
focused on obtaining theoretical results for specific types of graphs and specific
problem cases. Surprisingly, not a single metaheuristic approach has been applied
so far—to our knowledge—in order to tackle the problem from a practical point
of view. This was our main motivation for the development of an ant colony
optimization algorithm, and a hybrid variant of the algorithm that makes use
of a mathematical programming solver. In the following we provide a technical
description of the firefighter problem, as well as an overview on related work.

---

* This work was supported by grant TIN2012-37930 of the Spanish Government, and
project 2009-SGR1137 of the Generalitat de Catalunya.

C. Blum and G. Ochoa (Eds.): EvoCOP 2014, LNCS 8600, pp. 218–229, 2014.
© Springer-Verlag Berlin Heidelberg 2014

Problem Description. Given is an undirected graph $G = (V, E)$. Each vertex of the graph is initally labelled as *untouched*. At time $t = 0$, the fire breaks out at a pre-defined set $B_{\text{init}} \subseteq V$ of vertices. These vertices are labelled as *burnt*. Then, at time step $t = 1$, each member of a set of D firefighters (where D is a fixed input parameter of the problem) may choose a vertex from G. These D vertices are then labelled as *defended*. Moreover, these vertices may, of course, only be chosen from the set of vertices that are still labelled *untouched*. Time step $t = 1$ finishes by the fire propagating from the vertices labelled as *burnt* to all neighboring vertices that are labelled *untouched*. All these vertices change their label to *burnt*. This process proceeds ($t = 2, 3, \ldots$) until the fire cannot spread any further, that is, until the fire is contained. The optimization objective considered in this paper concerns the choice of the D defended vertices at each time step such that, when the fire is contained, the number of saved vertices is maximal. Note that a vertex is called saved if it is either labelled *defended* or *untouched* once the fire is contained.

In [11] was proposed a linear integer programming model of this problem. The model is based on two sets of binary variables. The first set consists of a binary variable $b_{v,t}$ for each vertex $v \in V$ and each time step $0 \le t \le T$, where T is an upper bound for the fire containment process, which must be appropriately set. A setting of $b_{v,t} = 1$ means that vertex v is labelled *burnt* at time step t, while the opposite is the case otherwise. The second set of variables contains a binary variable $d_{v,t}$ for each vertex $v \in V$ and each time step $0 \le t \le T$. A setting of $d_{v,t} = 1$ means that vertex v is labelled *defended* at time step t, while the opposite is the case otherwise. The model may then be stated as follows.

$$\max |V| - \sum_{v \in V} b_{v,T} \tag{1}$$

subject to:

$$b_{v,t} + d_{v,t} - b_{v',t-1} \ge 0 \quad \text{for } v \in V, v' \in N(v), \text{and } 1 \le t \le T \tag{2}$$

$$b_{v,t} + d_{v,t} \le 1 \quad \text{for } v \in V \text{and } 1 \le t \le T \tag{3}$$

$$b_{v,t} - b_{v,t-1} \ge 0 \quad \text{for } v \in V \text{and } 1 \le t \le T \tag{4}$$

$$d_{v,t} - d_{v,t-1} \ge 0 \quad \text{for } v \in V \text{and } 1 \le t \le T \tag{5}$$

$$\sum_{v \in V} (d_{v,t} - d_{v,t-1}) \le D \quad \text{for } 1 \le t \le T \tag{6}$$

$$b_{v,0} = 1 \quad \text{for } v \in B_{\text{init}} \tag{7}$$

$$b_{v,0} = 0 \quad \text{for } v \in V \setminus B_{\text{init}} \tag{8}$$

$$d_{v,0} = 0 \quad \text{for } v \in V \tag{9}$$

$$b_{v,t}, d_{v,t} \in \{0, 1\} \quad \text{for } v \in V \text{and } 1 \le t \le T$$

Hereby, constraints (2) ensure the spread of the fire while respecting defended vertices. Note, in this context, that the notation $N(v)$ refers to the neighboring vertices of a vertex $v \in V$. Constraints (3) prevent a firefighter from defending a burnt vertex and the fire from burning a defended vertex. Furthermore, constraints (4) ensure that a burnt vertex remains burnt, while constraints (5) ensure that a defended vertex remains defended. Finally, constraints (6) limit the number of firefighters per time step to D, and constraints (7-9) fix the initial conditions for time step $t = 0$, that is, the vertices from B_{init} are declared as burnt, and none of the vertices is declared as defended.

Related Work. The existing literature on the firefighter problem is focused on theoretical analysis [15]. The decision variant of the problem, saving just one vertex per time unit, was proved NP-complete for bipartite graphs in 2003 [25]. Stronger results appeared afterwards for cubic graphs [24] and graphs with degree three [14]. More recently, Cygan et al. analyzed the complexity of different parameterized versions of the problem on general graphs [10], and Bazgan et al. [3] and Costa et al. [9] analyzed the case with more than one firefighter.

In 2000, the greedy algorithm for trees, which saves the vertex v that maximizes the number of vertices that will be saved if v is protected, was proved to be a 1/2-approximation algorithm [22]. A linear programming relaxation for trees that supposedly gives a c-approximation algorithm was presented in 2006 [20], and a subexponential $(1 - 1/e)$-approximation method in 2008 [7]. These results have been improved in 2011 [23]. On the other hand, exact polynomial solutions exist for caterpillar and P-trees [15,19,25].

First approaches for grids of dimension 2 and 3 were provided in 2002 [17,28], and then generalized in 2007 [11]. These studies concluded that two firefighter were needed to contain the fire in an infinite 2-dimensional square grid, and $2d-1$ in a d-dimensional one with $d \geq 3$. There exist specific results for triangular, strong, and hexagonal grids [17,26,27], and for other graph classes [18].

The *surviving rate* of a graph is defined as the average percentage of vertices that can be saved when f fires break out at random vertices of the graph [6]. The study of this concept has become very fruitful in the literature and the evidence is the existence of many works on the subject for different graph structures (see [8,5,12], just to name a few).

Finally, variants of the firefighter problem are the fractional firefighter problem [17], the spreading vaccinations model [1,2,16], the non-constant firefighter problem [13,29], as well as other problem variants described in Section 8 of [15].

Outline of the Paper. The outline of the remaining part of the paper is as follows. The algorithms are described in Section 2. A description of the problem instance, the tuning process, and the experimental results are given in Section 3. Finally, conclusions and an outlook to future work are provided in Section 4.

2   The Proposed Algorithms

The algorithms proposed in this work for the firefighter problem are based on a $\mathcal{MAX}$–$\mathcal{MIN}$ Ant System ($\mathcal{MMAS}$) implemented in the Hyper-Cube Framework (HCF) [30,4]. In the following, we first provide a description of this pure approach, henceforth labelled ACO. Afterwards, a hybrid algorithm which combines the pure ACO approach with an integer programming solver is presented.

2.1   Solution Representation and Pheromone Model

Before being able to delve into the algorithm description, we must first deal with some basic aspects of the algorithm. A crucial aspect for any metaheuristic concerns the representation of valid solutions to the tackled problem. A valid solution in the case of the firefighter problem consists of the D vertices to be defended at each time step $t \geq 1$. Remember that only *untouched* vertices may be defended, that is, vertices that are neither *burnt* nor already *defended*. Note that any permutation π of all the vertices in V can be seen in the following way as a valid solution to the problem. In order to evaluate a solution given in the form of a permutation, the fire spreading process is simulated starting with the initial situation in which only the vertices from B_{init} are declared as *burnt*. For $t = 1$, the D vertices to be defended are derived by scanning π from left to right, starting at the left-most position. The scanning process accepts any vertex which is not already burnt. In case an already burnt vertex is encountered, the scanning process simply moves to the vertex at the next position to the right. At each time step t, the scanning process proceeds until D feasible vertices are found (or the end of the permutation is reached). Moreover, assuming that the scanning process has stopped at a position j of π at time step t, the scanning process for $t = 1$ is started at position $j + 1$. This process is continued as long as the fire may spread further. The output of the process—that is, the objective function value $f(\pi)$—is determined as the number of vertices that are not burnt.

A second crucial aspect, this time specific to ACO algorithms, concerns the definition of the pheromone model. We decided for the standard pheromone model which is used when solutions are represented as permutations. More specifically, the pheromone model $\mathcal{T}$ contains a pheromone value $\tau_{v,j}$ for each combination of a vertex v and a position $1 \leq j \leq |V|$ of a permutation.

2.2   ACO: A Pure $MMAS$ Approach

A $MMAS$ implemented in the HCF works as follows (see also Algorithm 1). First, n_a solutions are probabilistically generated, based on pheromone and greedy information. Second, the pheromone values are modified using (at most) three solutions: (1) the iteration-best solution π^{ib}, (2) the restart-best solution π^{rb}, and (3) the best-so-far solution π^{bs}. The general aim of the pheromone update is to focus the search on areas of the search space containing high-quality solutions. Moreover, the algorithm performs restarts—that is, re-initializations of the pheromone values—upon convergence. More specifically, restarts are controlled by the so-colled convergence factor (cf) and a Boolean control variable called *bs_update*. A detailed description of all algorithmic components is provided in the following.

Construct_Solution(): A solution π—that is, a permutation of all nodes of V—is probabilistically generated from left to right by simulating the fire spreading process just like in the case of a solution evaluation. The position counter i is initialized to 1. Moreover, in the initial situation (at $t = 0$) all vertices are

---

Algorithm 1. Pure ACO for the firefighter problem

1. **input:** an undirected graph $G = (V, E)$, a set B_{init}, and parameter D
2. $\pi^{bs} :=$ NULL, $\pi^{rb} :=$ NULL, $cf := 0$, $bs\_update :=$ **false**
3. $\tau_{v,j} := 0.5$ for all $\tau_{v,j} \in \mathcal{T}$
4. **while** termination conditions not met **do**
5.   **for** $cnt = 1, \ldots, n_a$ **do** $\pi_{cnt} :=$ Construct_Solution() **end for**
6. $\pi^{ib} := \text{argmin}\{f(\pi_{cnt}) \mid cnt = 1, \ldots, n_a\}$
7.   **if** $f(\pi^{ib}) > f(\pi^{rb})$ **then** $\pi^{rb} := \pi^{ib}$
8.   **if** $f(\pi^{ib}) > f(\pi^{bs})$ **then** $\pi^{bs} := \pi^{ib}$
9.   ApplyPheromoneUpdate($cf, bs\_update, \mathcal{T}, \pi^{ib}, \pi^{rb}, \pi^{bs}$)
10. $cf :=$ ComputeConvergenceFactor($\mathcal{T}$)
11.   **if** $cf > 0.99$ **then**
12.   **if** $bs\_update =$ **true then**
13. $\tau_{v,j} := 0.5$ for all $\tau_{v,j} \in \mathcal{T}$, $\pi^{rb} :=$ NULL, and $bs\_update :=$ **false**
14.   **else**
15. $bs\_update :=$ **true**
16.   **end if**
17.   **end if**
18. **end while**
19. **output:** π^{bs}, the best solution found by the algorithm

---

labelled *untouched*, apart from the vertices in B_{init} which are declared as *burnt*. For each $t > 1$, at most D vertices are chosen from the set of untouched vertices. Let us henceforth refer to this set as $U \subset V$. The probability to choose a vertex $v \in U$ is defined as follows:

$$\mathbf{p}(v) := \frac{\tau_{v,i} \cdot \eta_{v,i}}{\sum_{v' \in U} \tau_{v',i} \cdot \eta_{v',i}} \ , \tag{10}$$

where the pheromone value $\tau_{v,i}$ indicates the desirability to place vertex v at position i of solution/permuation π, while $\eta_{v,i}$ is the heuristic information for doing so. This heuristic information is defined as follows. Let π_η be the sequence in which the vertices of V are sorted increasingly according to their distance (in terms of the length of the shortest path) to the closest vertex from B_{init}. The degree of the vertices is hereby used to break ties, that is, vertices with higher degrees are preferred. Furthermore, the position of a vertex $v \in V$ in π_η is henceforth denoted by $\text{pos}(v, \pi_\eta)$. The heuristic information $\eta_{v,i}$ to place a vertex v at position i of a solution π is then defined as:

$$\eta_{v,i} := |V| - \frac{|i - \text{pos}(v, \pi_{eta})|}{|V|} \tag{11}$$

In words, the heuristic information for a vertex v (in conjunction with the current position i) decreases with increasing distance from $\text{pos}(v, \pi_\eta)$. Given the probabilities from Eqn. (10), a vertex $v \in U$ is chosen in the following way. First a value $0 \le r \le 1$ is drawn uniformly at random. In case $r \le q_0$, the vertex with the highest probability is chosen deterministically. Otherwise, a vertex is chosen

Table 1. Setting of κ_{ib}, κ_{rb}, and κ_{bs} depending on the convergence factor cf and the Boolean control variable bs_update

	$bs\_update = \text{FALSE}$				bs_update = TRUE
	$cf < 0.4$	$cf \in [0.4, 0.6)$	$cf \in [0.6, 0.8)$	$cf \geq 0.8$	
κ_{ib}	1	2/3	1/3	0	0
κ_{rb}	0	1/3	2/3	1	0
κ_{bs}	0	0	0	0	1

randomly according to the probabilities. Hereby, the *determinism rate* $q_0 \leq 1$ is a parameter of the algorithm.

Upon choosing a vertex v with respect to the probabilities shown in Eqn. (10), v is placed at position i of π, the position counter i is incremented, and v is declared as *defended*. After having selected D vertices in this way, t is incremented and the fire propagates from burnt nodes to all neighboring nodes which are still untouched. This process stops once the fire is contained. In a last step, π is completed by adding all vertices that are burnt or untouched in an arbitrary order to the free positions of π.

ApplyPheromoneUpdate(cf,bs_update,$\mathcal{T}$,π^{ib},π^{rb},π^{bs}): The three solutions π^{ib}, π^{rb}, and π^{bs} (as described at the beginning of this section) are used for the pheromone update. Their influence on the pheromone update depends on the current value of the convergence factor cf and on the value of the Boolean control variable bs_update as outlined in Table 1. Each pheromone value $\tau_{v,j} \in \mathcal{T}$ is updated as follows: $\tau_{v,j} := \tau_{v,j} + \rho \cdot (\xi_{v,j} - \tau_{v,j})$, where $\xi_{v,j} := \kappa_{ib} \cdot \Delta(\pi^{ib}, v, j) + \kappa_{rb} \cdot \Delta(\pi^{rb}, v, j) + \kappa_{bs} \cdot \Delta(\pi^{bs}, v, j)$. Hereby, κ_{ib} is the weight of solution π^{ib}, κ_{rb} the one of solution π^{rb}, and κ_{bs} the one of solution π^{bs}. Moreover, $\Delta(\pi, v, j)$ evaluates to 1 if and only if vertex v is at position j in solution π. Otherwise, the function evaluates to 0. Note also that the three weights must be chosen such that $\kappa_{ib} + \kappa_{rb} + \kappa_{bs} = 1$. After the pheromone update, pheromone values that exceed $\tau_{\max} = 0.99$ are set back to $\tau_{\max}$, and pheromone values that have fallen below $\tau_{\min} = 0.01$ are set back to $\tau_{\min}$. This prevents the algorithm from reaching the state of convergence.

ComputeConvergenceFactor($\mathcal{T}$): The convergence factor cf is computed on the basis of the pheromone values (see also [4]):

$$cf := 2 \left(\left(\frac{\sum\limits_{\tau_{v,j} \in \mathcal{T}} \max\{\tau_{\max} - \tau_{v,j}, \tau_{v,j} - \tau_{\min}\}}{|\mathcal{T}| \cdot (\tau_{\max} - \tau_{\min})} \right) - 0.5 \right)$$

This results in $cf = 0$ when all pheromone values are set to 0.5. On the other side, when all pheromone values have either value $\tau_{\min}$ or $\tau_{\max}$, then $cf = 1$. In all other cases, cf has a value between 0 and 1. This completes the description of all components of the proposed algorithm.

2.3 HyACO: A Hybrid ACO Variant

The hybrid algorithm that we propose in this paper is based on a combination of the pure ACO approach—as outlined in the previous section—and mathematical programming. More specifically, the hybrid algorithm (henceforth labelled HyACO) is a sequential approach that works as follows. First, pure ACO is applied with a certain computation time limit. Then, the best solution obtained by pure ACO is given to CPLEX as a starting solution (see, for example, page 531 of the IBM ILOG CPLEX V12.1 user manual: *Starting from a Solution: MIP Starts*). CPLEX is then used to apply *solution polishing* for a certain amount of CPU time to the given starting solution (see, for example, page 521 of the IBM ILOG CPLEX V12.1 user manual: *Solution polishing*). Solution polishing can be seen as a black box local search based on branch & cut, with the aim to improve a solution, rather than proving optimality. The best solution found after this phase is provided by HyACO as output.

3 Experimental Evaluation

We implemented the proposed algorithms in ANSI C++ using GCC 4.7.3 for compiling the software. Moreover, the mathematical program outlined in Section 1 was solved with IBM ILOG CPLEX V12.1. The same version of CPLEX was used within HyACO. The experimental results that we outline in the following were obtained on a cluster of PCs with "Intel(R) Xeon(R) CPU 5130" CPUs of 4 nuclii of 2 GHz and 4 Gigabyte of RAM.

Problem Instances. A random graph may be generated by choosing a fixed number of vertices (n) and a so-called *edge probability* (p_e). In the process of generating a random graph $G = (V, E)$ with n vertices, the edge probability p_e is used to determine for each possible edge $e = (v, v')$ between two vertices $v \in V$ and $v' \in V$, if e is added to E or not. For the purpose of evaluating the proposed algorithms, the following benchmark set of random graphs was generated. The number of vertices n was chosen to be from $\{50, 100, 500, 1000\}$. In the case of $n = 50$, edge probabilities $\{0.1, 0.15, 0.2\}$ were considered. In the case of $n = 100$, we considered edge probabilities $\{0.05, 0.075, 0.1\}$. In the case of $n = 500$, edge probabilities from $\{0.015, 0.02, 0.025\}$ were considered. Finally, in the case of $n = 1000$, we made use of edge probabilities $\{0.0075, 0.01, 0.0125\}$. For each combination of n and p_e we generated 10 random graphs, that is, in total 120 random graphs were generated. Note that different edge probabilities were considered in order to generate graphs of different densities. The average number of edges for the 10 graphs of each combination of n and p_e is shown in Table 2.

Parameter Setting and Tuning. In general, we assumed a standard parameter setting for the ACO algorithm. More specifically, the number of ants (n_a) was set to 10 per iteration, and the learning rate (ρ) was set to 0.1. Moreover, all tests carried out in this paper assume that B_{init}, the set of vertices in which the fire

Table 2. Average number of edges for each combination of n and p_e. The table contains the before-mentioned combinations in the form (n, p_e). The average number of edges is indicated underneath.

$(50, 0.1)$	$(50, 0.15)$	$(50, 0.2)$	$(100, 0.05)$	$(100, 0.075)$	$(100, 0.1)$
122.6	180.0	245.5	249.6	373.6	495.7

$(500, 0.015)$	$(500, 0.02)$	$(500, 0.025)$	$(1000, 0.0075)$	$(1000, 0.01)$	$(1000, 0.0125)$
1865.0	2470.7	3128.3	3753.0	5003.6	6248.2

breaks out, only contains the vertex with index 0. The most crucial parameter of the proposed ACO algorithm is the determinism rate q_0, which determines the *greediness* of the solution construction process. Therefore, this parameter was chosen to be subject of tuning. For this purpose we selected for each combination of n and p_e the first one of the 10 random graphs and applied the pure ACO algorithm for $q_0 \in \{0.0, 0.1, 0.2, 0.3, 0.4, 0.5, 0.6, 0.7, 0.8, 0.9\}$ exactly once to each of the selected graphs. As a computation time limit we used $n/2$ CUP seconds. Moreover, all runs were performed for different values of D, which is the upper bound of vertices that may be saved at each iteration of the fire propagation process. More specifically, we used $D \in \{1, \ldots, 10\}$. The results for each combination of a graph and a value for D were ranked. These ranks are shown in Figure 1, in the following way. The graphic contains a grid of barplots. The rows in this grid refer to different settings of D, and the columns refer to the edge probabilities used to generate the correspondings graphs. Hereby, *sparse* refers to the smallest one of the three edge probabilities for each n, *medium* to the second one, and *dense* to the largest one. The height of the bars of each barplot indicates the average rank of the experiments. The leftmost bar corresponds to the experiments with $q_0 = 0.0$, and the rightmost bar corresponds to the experiments with $q_0 = 0.9$. In general, the results indicate that when D is small the determinism rate should not be too high. However, with growing D the value for the determinism rate should grow. As a compromise we chose the following setting for q_0 for the final experiments. For $D \in \{1, 2, 3\}$, q_0 is set to 0.5. For $D \in \{4, 5, 6\}$, q_0 is set to 0.7, and for the rest of the values of D, q_0 is set to 0.9.

Results. ACO and HyACO were both applied exactly once to each of the 120 graphs of the benchmark set. The computation time limit for ACO was chosen to be $n/2$. The computation time limit for the first phase of HyACO (in which the pure ACO is applied) was chosen to be $n/4$, and the computation time limit for the second phase of HyACO (in which solution polishing is applied by CPLEX) was equally chosen to be $n/4$. In this way, the computation time limits for ACO and HyACO are exactly the same. Finally, CPLEX was used to solve the mathematical program with the same computation time limit as the one used for ACO and HyACO.

The results are provided in Table 3, separated into four subtables corresponding to graph size. Each subtable presents the results in 10 rows, one for each of the 10 possible values of D. For each algorithm is shown the average result over the corresponding 10 random graphs. For example, the average result of CPLEX for graphs with $n = 50$ and $p_e = 0.1$ (when $D = 1$) is 7.4. Remember that this

Table 3. Numerical results of ACO, HyACO, and CPLEX

(a) Results for graphs with 50 vertices.

D	Edge probability $p_e = 0.1$			Edge probability $p_e = 0.15$			Edge probability $p_e = 0.2$		
	CPLEX	**ACO**	**HyACO**	**CPLEX**	**ACO**	**HyACO**	**CPLEX**	**ACO**	**HyACO**
1	**7.4** (10/10)	**7.4**	**7.4**	**4.5** (10/10)	**4.5**	**4.5**	**3.1** (10/10)	**3.1**	**3.1**
2	**26.6** (10/10)	26.4	26.5	**9.7** (10/10)	**9.7**	**9.7**	**7.2** (10/10)	**7.2**	**7.2**
3	**41.8** (10/10)	40.9	41.6	**18.8** (10/10)	16.5	18.5	**11.2** (10/10)	11.1	**11.2**
4	**47.9** (10/10)	47.8	**47.9**	**31.2** (10/10)	30.5	30.9	**17.5** (10/10)	16	17.2
5	**48.5** (10/10)	**48.5**	**48.5**	**39.1** (10/10)	36.1	**39.1**	**27.7** (10/10)	26.1	27.6
6	**48.8** (10/10)	**48.8**	**48.8**	**43.7** (10/10)	42.7	**43.7**	**33** (10/10)	31.4	**33**
7	**49** (10/10)	**49**	**49**	**46.3** (10/10)	45.4	**46.3**	**37.5** (10/10)	35.7	**37.5**
8	**49** (10/10)	**49**	**49**	**48.1** (10/10)	46.8	**48.1**	**42.7** (10/10)	40.3	42.6
9	**49** (10/10)	**49**	**49**	**48.6** (10/10)	48.2	**48.6**	**46.1** (10/10)	44.4	**46.1**
10	**49** (10/10)	**49**	**49**	**48.8** (10/10)	**48.8**	**48.8**	**47.5** (10/10)	47.1	**47.5**

(b) Results for graphs with 100 vertices.

D	Edge probability $p_e = 0.05$			Edge probability $p_e = 0.075$			Edge probability $p_e = 0.1$		
	CPLEX	**ACO**	**HyACO**	**CPLEX**	**ACO**	**HyACO**	**CPLEX**	**ACO**	**HyACO**
1	**9.2** (10/10)	9.1	**9.2**	**5.4** (10/10)	**5.4**	**5.4**	**3.9** (10/10)	**3.9**	**3.9**
2	26.9 (4/10)	25.7	**27.6**	**11.3** (2/10)	11.2	**11.3**	8.5 (5/10)	8.3	**8.7**
3	**62.8** (5/10)	54.6	62.7	41.5 (3/10)	41	**41.6**	**21.4** (1/10)	21	21.3
4	85.3 (8/10)	66.3	**85.5**	**53.7** (4/10)	52.4	53.3	**25.5** (1/10)	24.5	**25.5**
5	**97.3** (10/10)	92.3	**97.3**	65.7 (5/10)	63.5	**65.9**	**30.2** (1/10)	29.1	29.5
6	**98.5** (10/10)	98.3	**98.5**	87.5 (8/10)	75.1	87.3	41.8 (2/10)	33.9	41
7	**98.8** (10/10)	**98.8**	**98.8**	**98.1** (10/10)	87.9	**98.1**	58.7 (3/10)	46.4	56.3
8	**98.9** (10/10)	**98.9**	**98.9**	**98.6** (10/10)	93.5	**98.6**	74.8 (6/10)	62	74
9	**99** (10/10)	**99**	**99**	**98.8** (10/10)	**98.8**	**98.8**	89.2 (8/10)	77.3	88
10	**99** (10/10)	**99**	**99**	**99** (10/10)	**99**	**99**	94.7 (9/10)	85.9	94.4

(c) Results for graphs with 500 vertices.

D	Edge probability $p_e = 0.015$			Edge probability $p_e = 0.02$			Edge probability $p_e = 0.025$		
	CPLEX	**ACO**	**HyACO**	**CPLEX**	**ACO**	**HyACO**	**CPLEX**	**ACO**	**HyACO**
1	7.6 (0/10)	7.5	**7.8**	5.3 (0/10)	5.2	**5.6**	4.2 (0/10)	4.3	**4.4**
2	5.6 (0/10)	13	**13.6**	10.6 (0/10)	10.4	**11.2**	**9.1** (0/10)	8.5	9
3	3.1 (0/10)	18.8	**21.3**	60.3 (1/10)	63.1	**63.6**	12.8 (0/10)	12.7	**13.8**
4	150.2 (3/10)	119.9	**168.3**	69.5 (1/10)	67.6	**70.4**	17.7 (0/10)	16.9	**18.6**
5	250.5 (5/10)	218.9	**265.9**	54.6 (1/10)	72.7	**74.6**	6.5 (0/10)	21.6	**22.4**
6	349.1 (7/10)	268.8	**363**	102.4 (2/10)	123.8	**126.5**	25.6 (0/10)	26.3	**33.8**
7	448.9 (9/10)	407.6	**453.5**	102.7 (2/10)	128.1	**130.1**	78.1 (1/10)	77.6	**93**
8	449.1 (9/10)	453.9	**455**	299 (6/10)	135.8	**315.5**	154.2 (3/10)	127.6	**173.5**
9	449.1 (9/10)	454.7	**456.6**	349 (7/10)	317.5	**363.8**	223.3 (3/10)	221.8	**225.4**
10	**498.8** (10/10)	455.5	**498.8**	409.2 (8/10)	321.1	**410.2**	221.5 (4/10)	225.1	**229.6**

(d) Results for graphs with 1000 vertices.

D	Edge probability $p_e = 0.0075$			Edge probability $p_e = 0.01$			Edge probability $p_e = 0.0125$		
	CPLEX	**ACO**	**HyACO**	**CPLEX**	**ACO**	**HyACO**	**CPLEX**	**ACO**	**HyACO**
1	105.2 (1/10)	107.4	**107.8**	4.9 (0/10)	5.7	**6**	4 (0/10)	**4.9**	4.7
2	107.9 (1/10)	112.7	**115**	4.4 (0/10)	**10.9**	10.6	8 (0/10)	9.4	**10.1**
3	101.7 (1/10)	**118**	**118**	13.7 (0/10)	15.9	**17**	4.6 (0/10)	**14.3**	13.9
4	399.4 (4/10)	318.3	**415.7**	14.8 (0/10)	21.4	**24.1**	99.9 (1/10)	**116.6**	116.5
5	399.6 (4/10)	419	**421.2**	104.3 (1/10)	27.1	**123.6**	103.3 (1/10)	120.8	**120.9**
6	598.8 (6/10)	423.6	**614.2**	201.5 (2/10)	129.1	**226**	99.9 (1/10)	125.5	**126.2**
7	898.1 (9/10)	523.5	**902.5**	299.7 (3/10)	**325.5**	325.2	199.8 (2/10)	**226.7**	226.6
8	**998.2** (10/10)	528.6	**998.2**	399.5 (4/10)	329.7	**424.4**	218.7 (2/10)	231.1	**234.8**
9	**998.9** (10/10)	905.6	**998.9**	399.6 (4/10)	**427.9**	427.7	301 (3/10)	**236.2**	332.1
10	**999** (10/10)	**999**	**999**	499.4 (5/10)	432.6	**528.4**	602.2 (6/10)	335.3	**620.5**

Fig. 1. Tuning results. The meaning of the barplots is outlined in the text.

value refers to the average number of vertices that could be saved from being burnt in the 10 corresponding random graphs. Therefore, the higher this value the better. The best result obtained by CPLEX, ACO and HyACO is always indicated in bold font. Moreover, in the case of CPLEX additional information is given in the form (x/y). Hereby, x indicates the number of cases (out of $y = 10$) that CPLEX could provenly solve to optimality within the given CPU time limit.

Several observations can be made. First, CPLEX is able to solve all cases with $n = 50$ to optimality. Not surprisingly, the difficulty for CPLEX grows with increasing graph size and increasing graph density. Moreover, the difficulty for CPLEX seems to decrease with increasing values of D. Concerning the ACO algorithms we can observe that the hybrid version (HyACO) is clearly superior to the pure ACO version. Concerning the comparison to CPLEX we can state that the pure ACO version is clearly inferior to CPLEX for what concerns rather small graphs ($n \in \{50, 100\}$). However, with growing graph size and graph density ACO achieves to outperform CPLEX especially for smaller values of D (observe, for example, the results for $n = 1000$ and $p_e = 0.0125$). The hybrid ACO variant, on the other side, generally outperforms both CPLEX and ACO when $n \in \{500, 1000\}$. Moreover, concerning the graphs with $n \in \{50, 100\}$, HyACO often matches the results of CPLEX, and is only slightly inferior in the remaining cases. In summary, we can say that CPLEX is the recommended solution method for rather small graphs, while HyACO is the recommended solution method when graph size and density increase.

4   Conclucions and Future Work

In this work we presented the first metaheuristics for the so-called firefighter problem. In particular, we proposed a pure ACO approach and a hybrid ACO variant that makes use of a mathematical programming solver (in our case: CPLEX). The results show that CPLEX obtains the best results for rather small graphs, while the hybrid technique is superior to CPLEX and the pure ACO approach for larger graphs.

The main line of research for future work concerns the application of the proposed algorithms to different families of graphs, such as, for example, random geometric graphs and small world networks.

References

1. Anshelevich, E., Chakrabarty, D., Hate, A., Swamy, C.: Approximation Algorithms for the Firefighter Problem: Cuts over Time and Submodularity. In: Dong, Y., Du, D.-Z., Ibarra, O. (eds.) ISAAC 2009. LNCS, vol. 5878, pp. 974–983. Springer, Heidelberg (2009)
2. Anshelevich, E., Chakrabarty, D., Hate, A., Swamy, C.: Approximability of the Firefighter Problem. Algorithmica 62(1-2), 520–536 (2010)
3. Bazgan, C., Chopin, M., Ries, B.: The firefighter problem with more than one firefighter on trees. Discrete Applied Mathematics 161(7-8), 899–908 (2013)
4. Blum, C., Dorigo, M.: The hyper-cube framework for ant colony optimization. IEEE Trans. on Man, Systems and Cybernetics – Part B 34(2), 1161–1172 (2004)
5. Bonato, A., Messinger, M.E., Prałat, P.: Fighting constrained fires in graphs. Theoretical Computer Science 434, 11–22 (2012)
6. Cai, L., Cheng, Y., Verbin, E., Zhou, Y.: Surviving Rates of Graphs with Bounded Treewidth for the Firefighter Problem. SIAM Journal on Discrete Mathematics 24(4), 1322–1335 (2010)
7. Cai, L., Verbin, E., Yang, L.: Firefighting on Trees $(1 - 1/e)$–Approximation, Fixed Parameter Tractability and a Subexponential Algorithm. In: Hong, S.-H., Nagamochi, H., Fukunaga, T. (eds.) ISAAC 2008. LNCS, vol. 5369, pp. 258–269. Springer, Heidelberg (2008)
8. Cai, L., Wang, W.: The Surviving Rate of a Graph for the Firefighter Problem. SIAM Journal on Discrete Mathematics 23(4), 1814–1826 (2010)

9. Costa, V., Dantas, S., Dourado, M.C., Penso, L., Rautenbach, D.: More fires and more fighters. Discrete Applied Mathematics 161(16-17), 2410–2419 (2013)
10. Cygan, M., Fomin, F.V., van Leeuwen, E.J.: Parameterized Complexity of Fire-fighting Revisited. In: Marx, D., Rossmanith, P. (eds.) IPEC 2011. LNCS, vol. 7112, pp. 13–26. Springer, Heidelberg (2012)
11. Develin, M., Hartke, S.G.: Fire containment in grids of dimension three and higher. Discrete Applied Mathematics 155(17), 2257–2268 (2007)
12. Esperet, L., van den Heuvel, J., Maffray, F., Sipma, F.: Fire Containment in Planar Graphs. Journal of Graph Theory 73(3), 267–279 (2013)
13. Feldheim, O.N., Hod, R.: 3/2 Firefighters Are Not Enough. Discrete Applied Mathematics 161(1-2), 301–306 (2013)
14. Finbow, S., King, A., MacGillivray, G., Rizzi, R.: The firefighter problem for graphs of maximum degree three. Discrete Mathematics 307(16), 2094–2105 (2007)
15. Finbow, S., Science, C., Scotia, N., Macgillivray, G.: The Firefighter Problem: A survey of results, directions and questions. Australian Journal of Combinatorics 43, 57–77 (2009)
16. Floderus, P., Lingas, A., Persson, M.: Towards more efficient infection and fire fight-ing. In: CATS 2011 Proceedings of the Seventeenth Computing: The Australasian Theory Symposium, pp. 69–74 (2011)
17. Fogarty, P.: Catching the fire on grids. Master's thesis, Department of Mathematics. University of Vermont, USA (2003)
18. Fomin, F.V., Heggernes, P., van Leeuwen, E.J.: Making life easier for firefighters. In: Kranakis, E., Krizanc, D., Luccio, F. (eds.) FUN 2012. LNCS, vol. 7288, pp. 177–188. Springer, Heidelberg (2012)
19. Grötschel, M., Lovász, L., Schrijver, A.: Geometric Algorithms and Combinatorial Optimization. Springer (1988)
20. Hartke, S.G.: Attempting to Narrow the Integrality Gap for the Firefighter Problem on Trees. In: DIMACS Series in Discrete Mathematics and Theoretical Computer Science, pp. 225–231 (2006)
21. Hartnell, B.: Firefighter! An application of domination. In: 20th Conference on Numerical Mathematics and Computing (1995)
22. Hartnell, B., Li, Q.: Firefighting on trees: How bad is the greedy algorithm? In: Proc. of the Thirty-first Southeastern International Conference on Combinatorics, Graph Theory and Computing, pp. 187–192 (2000)
23. Iwaikawa, Y., Kamiyama, N., Matsui, T.: Improved Approximation Algorithms for Firefighter Problem on Trees. IEICE Transactions on Information and Systems E94-D(2), 196–199 (2011)
24. King, A., MacGillivray, G.: The firefighter problem for cubic graphs. Discrete Mathematics 310(3), 614–621 (2010)
25. MacGillivray, G., Wang, P.: On the firefighter problem. Journal of Combinatorial Mathematics and Combinatorial Computing 47, 83–96 (2003)
26. Messinger, M.E., Scotia, N.: Firefighting on the Triangular Grid. Journal of Combinatorial Mathematics and Combinatorial Computing 63, 3–45 (2007)
27. Messinger, M.E.: Firefighting on Infinite Grids. Master's thesis, Department of Mathematics and Statistics, Dalhousie University, Halifax, Canada (2004)
28. Moeller, S., Wang, P.: Fire Control on graphs. Journal of Combinatorial Mathematics and Combinatorial Computing 41, 19–34 (2002)
29. Ng, K., Raff, P.: A generalization of the firefighter problem on. Discrete Applied Mathematics 156(5), 730–745 (2008)
30. Stützle, T., Hoos, H.H.: $\mathcal{MAX}$-$\mathcal{MIN}$ Ant System. Future Generation Computer Systems 16(8), 889–914 (2000)

The Influence of Correlated Objectives on Different Types of P-ACO Algorithms

Ruby L.V. Moritz*, Enrico Reich, Matthias Bernt, and Martin Middendorf

Parallel Computing and Complex Systems Group, Institute of Computer Science,
University of Leipzig, Germany
{ruby.moritz,bernt,middendorf}@informatik.uni-leipzig.de

Abstract. The influence of correlated objectives on different types of
P-ACO algorithms for solutions of multi objective optimization problems
is investigated. Therefore, a simple method to create multi objective optimization problems with correlated objectives is proposed. Theoretical
results show how certain correlations between the objectives can be obtained. The method is applied to the Traveling Salesperson problem. The
influence of the correlation type and strength on the optimization behavior of different P-ACO algorithms is analyzed empirically. A particular
focus is given on P-ACOs with ranking methods.

1 Introduction

In many practical multi objective optimization problems (MOPs) the different
objectives are not independent from each other but correlated (e.g. [1]). Thus, it
is interesting to investigate the influence of correlations between the objectives
on the performance of different metaheuristics or other types of search algorithms, e.g. local search. The performance of metaheuristics is typically tested
on randomly created problem instances or on instances from benchmark libraries.
Often in these cases the objectives are independent (in particular when random
instances are used) or the correlation has not been investigated. Some authors
have investigated the effect of correlations between the objectives of MOPs on
algorithms in recent years.

It has been pointed out in [2] that if the objectives of a MOP are not positively correlated their optima will, in general, be very different. This could be a
problem for genetic algorithms (GAs) because it suggests that recombination of
solutions that are good for different objectives is unlikely to yield good offspring.
In a preliminary study it was argued in [3] that correlation between objectives
might have an influence on the correlation between optima. It was shown in [4]
that the performance of local search operators for MOPs is strongly influenced
by the strength of the correlation between the objectives. In [5] it was shown

---

* This work has been supported by the European Social Fund (ESF) and the Free
  State of Saxony within Nachwuchsforschergruppe "Schwarm-inspirierte Verfahren
  zur Optimierung, Selbstorganisation und Ressourceneffizienz".

C. Blum and G. Ochoa (Eds.): EvoCOP 2014, LNCS 8600, pp. 230–241, 2014.

that the performance of hybrid algorithms might be less influenced by correlation effects. For ant colony optimization (ACO) algorithms the conclusion of [6] was that for highly positively correlated instances "less aggressive" ACO strategies (e.g. iteration-best instead of best-so-far pheromone update) perform better. On instances with weakly or negatively correlated objectives this does not hold. Garret et al. [7] studied the influence of correlation on the relative performance of two hybrid metaheuristics. Verel et al. [8] showed that if the objective correlation for NK-landscapes is negative the number of Pareto optimal solutions is large . In contrast, when the objective correlation is positive, the number of Pareto optimal solutions is small and a metaheuristic might be able to find a large fraction of them. The co-influence of objective correlation, objective space dimension, and the degree of non-linearity on the number of Pareto optimal solutions was investigated in [9]. Shi et al. [10] studied community detection in networks as a MOP. They divided the relations between any two objective functions into three categories: positively correlated, independent, and negatively correlated. Then they compared the performance of the GA NSGA-Net for pairs of objective functions from the different categories to a single objective optimization based approach optimizing the original single objectives. The results show that NSGA-Net remarkably improves the performance for pairs of negatively correlated objectives, but not with a pair of positively correlated or independent objectives.

Ishibuchi et al. [11,12,13] investigated the influence of correlated objectives on evolutionary multi objective algorithms (EMOs) for the multi objective 0/1 Knapsack problem. One conclusion was that the performance of NSGA-II and SPEA2 (MOEA/D) was not (respectively was) severely degraded by an increase in the number of objectives when they were highly correlated [11]. Hence, Pareto dominance-based EMOs which typically have problems with a large number of objectives can profit from positive correlation. Thus, the difficulty of MOPs depends on the specific EMO algorithm as well as on the correlation between the objectives. NSGA-II and SPEA2, for instance, worked well on MOPs with highly correlated objectives [12]. The search of the hypervolume-based SMS-EMOA is biased toward the region of the Pareto front with good values for duplicated objectives [13]. Several authors pointed out that positive correlations between objectives of MOPs could be used for dimensionality reduction which can be exploited by metaheuristics [14,15]. Correlation may also be used to define groups of objectives that can be aggregated [16].

In this paper we propose a simple method to create MOP instances with different types of correlations between the objectives. Theoretical results show how certain correlations between the objectives can be obtained with this method. The influence of different correlations between the objectives on Population-based ACO (P-ACO) algorithms [17] using different ranking methods for selecting good non dominated solutions is studied empirically for the multi objective Traveling Salesperson problem. In the next section we give a short overview on methods to create instances of MOPs with correlated objectives and propose our new generation method.

2 MOPs with Correlated Objectives

The first problem generator for MOPs with correlated objectives was proposed for the multi objective quadratic assignment problem (QAP) with m flow matrices that define the different objectives [3,18]. The problem instance generator starts with a given flow matrix, e.g. from a real world instance. The other flow matrices are then generated such that an entry in the jth matrix, $j \in [2 : k]$, is a random variable that is correlated with the corresponding value in the first matrix. The degree of correlation can be defined for the jth matrix by a parameter $\alpha_j \in [-1, 1]$. Furthermore, the ratio of random and correlated entries in the generated matrix may be set. Several authors have used the QAP generator to study the influence of correlation – mostly for bi-objective problems – on the optimization behavior of metaheuristics and local search operators [4,6,5,7,19].

The creation of instances of the multi objective Traveling Salesperson problem (TSP) with correlated objectives was covered by [20]. For each pair of cities (i, j) k distance values $d_h(i, j)$, $h \in [2 : k]$ were created as

$$d_h(i, j) = c \cdot d_{h-1}(i, j) + (1 - c) \cdot rand \tag{1}$$

where the values $d_1(i, j)$ are chosen uniformly at random from $[0, 1]$, $c \in [-1, 1]$ is a "correlation parameter", and $rand$ is a uniform random number from $[0, 1]$. Observe, that Eq. (1) proposed by [20] can be somewhat problematic: i) the distance values can become > 1 for $c < 0$ even for original distance values from $[0, 1]$ which might lead to an uneven influence of different objectives on the behavior of metaheuristics, ii) the distance values are randomized to a different extent for c and $-c$.

Verel et al. [8] recently argued that only very few benchmark sets take the correlation of objectives into account (and that to the best of their knowledge, the QAP should be the only one). Therefore, they proposed a method to design MOPs where the correlation between the objectives is defined by a correlation matrix. In particular, NK-landscapes have been investigated but the method can also be applied more generally (see [8]). For their empirical investigations the same correlation strength was used for each pair of objectives.

The influence of correlated objectives on EMOs was studied in [11,12,13] for the multi objective n-item 0/1 Knapsack problem. Given are n items, k knapsacks with capacity c_j, and for each pair (i, j) a weight w_{ij} and a profit p_{ij}, $i \in [1 : n]$, $j \in [1 : k]$. The problem is to find a solution $x = (x_1, \dots, x_k) \in \{0, 1\}^k$ such that $f = (f_1, \dots, f_k)$ is maximized where $f_j = \sum_{i=1}^{n} p_{ij} x_i$ is subject to the constraint that $\sum_{i=1}^{n} w_{ij} x_i \le c_j$ for $j \in [1 : k]$. For even k correlated objective functions $g = (g_1, \dots, g_k)$ have been defined as $g_1 = f_1$, $g_2 = f_2$, $g_{2h+1} = \alpha f_{2h+1} + (1 - \alpha) f_1$, and $g_{2h+2} = \alpha f_{2h+2} + (1 - \alpha) f_2$ for $h \in [1, k/2 - 1]$ and parameter $\alpha \in (0, 1)$. A second type of correlated objective functions (called depended) $g' = (g'_1, \dots, g'_{10})$ have been defined as $g'_1 = f_1$, $g'_2 = f_2$, $g'_3 = f_1 + \alpha f_2$, $g'_4 = f_2 + \alpha f_1$, $g'_5 = f_1 - \alpha f_2$, $g'_6 = f_2 - \alpha f_1$, $g'_7 = f_1 + \beta f_2$, $g'_8 = f_2 + \beta f_1$, $g'_9 = f_1 - \beta f_2$, $g'_{10} = f_2 - \beta f_1$ for parameters $\alpha, \beta \in (0, 1)$. An extreme case of correlated, i.e. duplicated, objectives has been investigated in [13] (the test problems were (f_1, f_2, f_2), (f_1, f_2, f_2, f_2), and $(f_1, f_2, f_2, f_2, f_2)$).

Most of the above mentioned problem generation methods ([20,11,3]) generate problem instances with an inhomogeneous correlation structure, i.e. different strengths of pairwise correlations occur between the objectives. In the method of [3] each of the objectives 2 to k has a defined correlation with objective 1. But the correlation between two of the objectives 2 to k might be different. In the method used in [20] objective h, $h \in [2 : k]$ has a defined correlation with objective $h - 1$. But there are many different pairwise correlations between objectives i and j with $|i - j| \geq 2$. Also, for each of the problems created by the methods of [11,12,13] different pairwise correlations occur. The only MOPs with homogeneous correlation structure, i.e. where all pairwise correlations between different objectives are equal, are the NK-landscape instances from [8]. Surprisingly homogeneous correlation structures have to the best of our knowledge not yet been investigated for application oriented MOPs.

We consider a homogeneous correlation structure as the most basic case for the investigation of the influence of correlations on certain properties of the fitness landscape and on the performance of metaheuristics. Therefore, we propose a simple method to create homogeneous or only slightly heterogeneous problem instances and apply it to the TSP. Given is a set of n cities C and k distance functions d_h such that $d_h(i, j)$ is the h-distance between cities i and j (or in other words is the h-length of the edge (i, j), $i \neq j$). The problem is to find a Hamiltonian path such that the total h-length of the path is minimal with respect to each objective h, i.e. the sum of the h-lengths of the edges on the path should be minimal for $h \in [1 : k]$.

The creation of a multi objective TSP instance is done here starting with a single objective TSP instance (a set of cities C and distance function d) which can be taken from a standard benchmark library, e.g. the TSPLIB, from an application scenario, or it can be created artificially. We assume here that all distance values are from $[0, 1]$ which can be achieved by normalization. Distance d is used to create k distance functions $d_1, \ldots, d_k$. For a correlation factor $c \in [-1, 1]$ the distance function d_h is defined by

$$d_h(i, j) = \begin{cases} c \cdot d(i, j) & +(1 - c) \cdot rand \text{ for } 0 \leq c \leq 1 \\ |c| \cdot (1 - d(i, j)) & +(1 - |c|) \cdot rand \text{ for } -1 \leq c < 0 \end{cases} \tag{2}$$

Observe, that this solves the two problems of Eq. (1).

3   Ranking Methods for MOPs and P-ACO

A common principle of many metaheuristics is to select the good solutions from a set of solutions. For MOPs the dominance criterion is often used for the selection. Formally, a MOP is defined by a set of solutions X and a vector of *objective functions* $\boldsymbol{f}(a) = (f_1(a), \ldots, f_d(a))$, $f_i : X \mapsto \mathbb{R}$. The aim is to find solutions from X that minimize the objectives, i.e. $\min_{a \in X} \boldsymbol{f}(a) = \min_{a \in X}(f_1(a), \ldots, f_d(a))$. For $a, b \in X$, solution a *dominates* b if $a \prec b \iff \forall i \in [1 : d] : f_i(a) \leq f_i(b) \land \boldsymbol{f}(a) \neq \boldsymbol{f}(b)$. A solution $a \in X$ is called *pareto optimum* if $\nexists b \in X : b \prec a$. A solution $a \in X$ is called *non dominated* solution with respect to a subset $X' \subset X$ if

Algorithm 1. Multi objective P-ACO with ranking

---

1 $P \leftarrow \{\}$;
2 Initialize pheromone matrix $[\tau_{ij}]$;
3 **repeat**
4      $L \leftarrow \{\}$;
5   **foreach** *of the l ants* **do**
6   Construct a solution S using pheromone matrix $[\tau_{ij}]$;
7          $L \leftarrow L \cup \{S\}$;
8   Sort the solutions in $L \cup P$ with a ranking method;
9   Let P be the set of p best solutions from $L \cup P$;
10   Compute a pheromone matrix $[\tau_{ij}]$ from P;
11 **until** t *iterations done*;
12 **return** *non dominated solutions in* $L \cup P$

$\nexists b \in X' : b \prec a$. The *pareto set* is the set of all pareto optimal solutions from X. The corresponding set of objective values in $\mathbb{R}^d$ is called the *pareto front* of X.

For MOPs with a large number of objectives metaheuristics often have the problem that the fraction of non dominated solutions within a set of solutions is typically very large. Then, the selection of good solutions cannot be solely based on the pareto dominance relation. Therefore, several ranking methods have been proposed for selecting the good solutions from a set of non dominated solutions (see [21,22,23] on their influence within a GA for MOPs with many objectives).

Several variants of the pareto dominance relation have been proposed (see [24] for an overview). Here relation *favour* is relevant [25]. It compares two solutions with respect to the number of objectives in which one solution is better than the other solution. Based on relation favour two relations have been introduced in [24] that are also considered here. The WL relation is based on the number of won comparisons and lost comparisons with respect to relation favour. For $A \subseteq X$ define $a \prec_{\mathrm{WL}}{}^A b$ iff either a is favoured over more solutions in A than b or a and b are each favoured over the same number of solutions in A and the number of solutions in A that are favoured over a is less or equal than the corresponding number for b.

Relation Points tries to solve a potential problem of relation WL, namely that it ignores ties. A scoring scheme is defined by the *point score* $S(a, A)$ of a solution $a \in X$ with respect to a set of solutions $A \subseteq X$ such that $S(a, A)$ is w times the number of solutions $x \in A$ where a is favoured over x plus the number of solutions $y \in A$ where neither a is favoured over y nor y over a. Parameter $w \geq 1$ is a constant that specifies how many points are attributed to a won comparison. The Points relation is then defined by $a \prec_{\mathrm{Pt}}{}^A b \iff S(a, A) \geq S(b, A)$. The favour, points, and WL relation are refinements of the pareto dominance relation [24]. It should be noted that other ranking methods exist (see [26,27] for an overview).

The P-ACO uses a population P of p solutions that is transferred from one iteration to the next [17]. The pheromone information is computed from the solutions that are currently in P. As in traditional ACO algorithms the ants use the pheromone information to compute a set of new solutions L. For the TSP with n cities the pheromone information is stored in a pheromone matrix τ_{ij}, $i, j \in [1 : n]$. The pheromone values are $\tau_{ij} = t_{init} + \delta \cdot k$ where τ_{init} and δ are parameters of the algorithm and k is the number of tours in P that have city j after city i. Based on the ranking relations a population update method was proposed [24] (Algorithm 1). The solutions in $L \cup P$ are sorted with a ranking method and the best p solutions form the population of the next iteration. It was shown empirically that the P-ACOs with the new ranking schemes can profit from the ranking methods when compared to a Crowding P-ACO [28].

For the experimental evaluation we also used two ranking methods that are based on a relatively direct evaluation of the objective values. The first method "Rel" is based on function $R(a, b) = \sum_{i=1}^{d} f_i(a)/f_i(b)$ for two solutions $a, b \in X$ where d is the number of objectives. Rel is defined by $a \prec_{\text{Rel}} b$ iff $R(a, b) < d$. The second method "W" uses the weighted sum of the objectives. Here, random weights w_i with $i \in [1 : d]$ and $\sum_{i=1}^{d} w_i = 1$ are used. The weights are chosen uniformly at random for each ant in each iteration anew and are averaged with those of the previous iteration to change the weights more smoothly. The score of a solution for the ranking is computed with the current weighting scheme of the ant that constructed the solution.

The P-ACOs with the ranking methods are compared to: i) the Standard P-ACO (STD-P-ACO) [29] and ii) the $Crowding$ P-ACO (CR-P-ACO) [28]. The STD-P-ACO computes in every iteration the non dominated front from the solutions in $P \cup L$. The new population for the next iteration consists of a randomly chosen solution s from the non dominated front plus the $p - 1$ solutions that are closest to s in the objective space. In the CR-P-ACO the amount of updated pheromone is inversely proportional to the dominance depth of the corresponding solution (see [30]). Population update is done by a crowding scheme: One of the newly generated solutions replaces the most similar solution of a randomly chosen subset of the current population if it dominates it. Similarity is measured in the solution space as the number of common adjacent cities.

All results are averages over 10 experiments. The same parameters as in [24] are used for P-ACO, i.e. each run was for $t = 2500$ iterations with population size $p = 5$ and $l = 25$ new solutions were created at every iteration. To improve the comparability with CR-P-ACO its population size and number of ants are set to $l = p = 25$ and the size of the random subsets is set to 5. Pheromone update of the P-ACOs is done with $\tau_{init} = 1/(n - 1)$ and $\delta = 1/(p + n)$. A combined non dominated front is computed as the union of the non dominated fronts of all runs. Since there is no standard heuristic for multiobjective TSP and to make our analysis no dependent on a specific heuristic the algorithms are used without heuristic.

The TSP test instances have $d = 5$ objectives. The single objective instances that were used to create the multi objective instances are symmetric TSP instances from the TSPLIB: ulysses22, att48, eil51, bier127, gr137, and rat195. Two types of multi objective test instances were created: i) homogeneous test instances: all objectives are positively correlated with the initial distance matrix

M and ii) heterogeneous test instances: only four objectives are positively correlated with M and the fifth objective is negatively correlated. For one test each objective was created with the same absolute correlation factor, i.e. c or $-c$. Different values of $|c| \in \{0.2, 0.4, 0.6, 0.8, 0.9\}$ were used.

4   Results

4.1   Expected Correlation

In the following we derive a formula for the expected Pearson correlation between the objectives for the method proposed in Section 2. We assume here that the initial problem is given in form of a matrix $M = [m_{ij}] \in [0:1]^{n_1 \times n_2}$. The random influence is given as an $n_1 \times n_2$ random matrix $R = [rand_{ij}] \in [0,1]^{n_1 \times n_2}$ where $rand_{ij}$ is chosen uniformly at random from $[0,1]$. In case of the TSP $n_1 = n_2 = n$, $M = [d_{ij}]$ is the distance matrix, and for R $rand_{ij}$ is chosen uniformly at random from $[0,1]$ for $i \neq j$ and $rand_{ii} = 0$. For a symmetric TSP additionally $d_{ij} = d_{ji}$ and $rand_{ij} = rand_{ji}$. $R \in [0:1]^{n_1 \times n_2}$ is uniformly distributed, i.e. expectation $E(R) = \frac{1}{2}$ and variance $Var(R) = \frac{1}{12}$. The distribution of $M \in [0:1]^{n_1 \times n_2}$ is unknown with the expected value $E(M)$ and variance $Var(M)$.

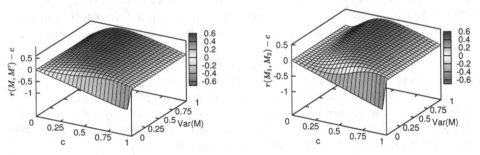

Fig. 1. Pearson correlation coefficient $r(M, M') - c$ (left) and $r(M_1, M_2) - c$ (right)

The generated problem matrix $M^+ = cM + (1-c)R \in [0:1]^{n_1 \times n_2}$ for the correlation factor $c \in [0:1]$ is positively correlated with M and $M^- = |c|(1-M) + (1-|c|)R \in [0:1]^{n_1 \times n_2}$ for $c \in [-1:0]$ is negatively correlated with M. To derive the expected correlation between two matrices the Pearson correlation coefficient r is applied It can be shown that for $M' \in \{M^+, M^-\}$ (due to limited space we omit the details of all computations in this section)

$$r(M, M') = \frac{\sum(m_{ij} - \overline{m})(m'_{ij} - \overline{m'})}{\sqrt{\sum(m_{ij} - \overline{m})^2 \sum(m'_{ij} - \overline{m'})^2}} = \frac{cVar(M)}{\sqrt{c^2 Var(M)^2 + \frac{(1-|c|)^2 Var(M)}{12}}}$$

(3)

The last equation can be obtained after deriving the following equations $E(M') = cE(M)+(1-c)/2$, $Var(M') = c^2 Var(M)+(1-c)^2/12$, $E(MM') = c(Var(M)+$

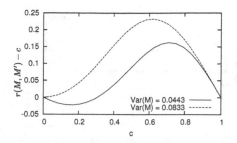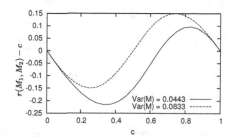

Fig. 2. Expected Pearson correlation coefficient $r(M, M') - c$ (left) and $r(M_1, M_2) - c$ (right) for two different variances of M

$E(M)^2) + (1 - c)E(M)/2$, and by the fact that $cov(M, R) = 0$. Fig. 1 (left) visualizes the difference of the expected Pearson correlation coefficient with c, i.e. $r(M, M') - c$, for different values of $Var(M)$ and $c \in [0, 1]$. For the special case that M is uniformly distributed in $[0, 1]$, i.e. $E(M) = 1/2$ and $Var(M) = 1/12$, this gives $r(M, M') = c/\sqrt{c^2 + (1 - c)^2}$. Fig. 2 (left) shows how the difference of the expected Pearson correlation coefficient with c, i.e. $r(M, M') - c$, changes with c for the variance of the test instances used and for uniformly at random distributed matrices with $Var(M) = 1/12 \approx 0.0833$. In order to obtain a specific Pearson correlation coefficient $r(M, M') = r$ the correlation factor c has to be chosen as $c = 1/(1 + 2\sqrt{3}\sqrt{(1/r - 1)Var(M)})$.

Now the question is, what is the Pearson correlation between different objectives for instances generated with the method proposed in Section 2, i.e. what is the correlation between two matrices that have been derived from M. We consider two cases. In the first case M_1 and M_2 have been generated from matrix M with the same correlation c. In the second case matrix M_1 is positively correlated with M and M_2 is negatively correlated with M. For the first case it can be shown that $r(M_1, M_2) = c^2 Var(M)/(c^2 Var(M) + (1 - |c|)^2/12)$. In the second case $r(M_1, M_2) = -c^2 Var(M)/(c^2 Var(M) + (1 - |c|)^2/12)$. Fig. 1 (right) shows $r(M, M') - c$ for the first case and different values of $Var(M)$ and c. Figure 2 (right) shows how $r(M_1, M_2) - c$ changes in the first case with c for two exemplary chosen variances for M. To obtain a specific Pearson correlation coefficient $r(M_1, M_2) = r$ for M_1 and M_2 the correlation factor c has to be chosen as $c = \pm 1/(1 + \sqrt{12 Var(M)(1 - r(M_1, M_2))/r(M_1, M_2)})$.

4.2 Experimental Results

In the first experiment it is shown that iterative correlation methods were objective i is correlated with objective $i - 1$ have the effect that a strong correlation occurs only for neighbored (with respect to the numbering) objectives. Fig. 3 shows the Pearson correlation that was measured between M and M_i when our method from Section 2 is applied iteratively, i.e. matrix M_i was created from matrix M_{i-1}, $i \geq 1$ and $M_0 = M$ as in [20]. Here M contains values chosen uniformly at random from $[0, 1]$. For $c < 0$ the average correlation between matrices

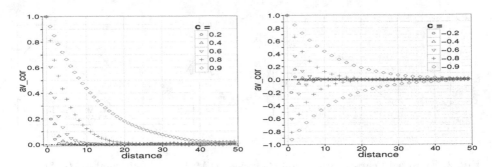

Fig. 3. Average Pearson correlation between matrix M_{i+d} and M_i for different distances d, positive correlation $c > 0$ (left) and negative correlation $c < 0$ (right)

Table 1. Average size of non dominated front and fraction of solutions in the common non dominated front; numbers are averaged over the five TSP instances; homogeneous case (upper part) heterogeneous case (bottom part)

$|c|$	average size						fraction (in %)					
	Pt	WL	W	Rel	CR	STD	Pt	WL	W	Rel	CR	STD
0.2	8.6	8.3	6.9	7.0	16.3	19.2	16.6	19.5	30.8	32.7	0.2	0.2
0.4	6.4	6.3	5.9	5.9	13.8	16.2	17.3	21.1	29.7	31.8	0.1	0.1
0.6	4.8	5.1	5.1	5.1	8.8	9.8	21.1	22.9	26.4	29.3	0.0	0.3
0.8	4.0	4.2	4.3	4.3	3.6	4.5	22.2	22.5	21.9	22.1	0.0	11.3
0.9	3.0	3.0	3.0	3.0	1.9	3.2	14.7	21.9	20.9	18.8	0.0	23.7
0.2	9.3	8.9	7.2	7.5	16.8	19.6	16.5	19.6	31.0	32.5	0.2	0.2
0.4	9.6	9.1	7.6	8.7	16.4	19.3	17.0	20.2	30.7	31.5	0.3	0.4
0.6	14.7	14.5	12.8	15.7	16.2	19.1	18.7	20.9	27.3	26.8	2.5	3.7
0.8	22.2	22.3	22.1	22.4	17.8	20.8	19.2	19.5	20.9	19.3	9.0	12.2
0.9	24.8	24.9	24.9	24.9	20.4	24.0	18.4	18.2	18.5	17.9	11.8	15.2

with uneven distance is negative. It can be seen that for small correlation values $|c| \leq 0.4$ the correlation between M_{i+5} and M_i is already close to zero. Hence, one has to be careful when observing that a metaheuristic is robust against correlated objectives for test instances of an iterative generation method.

Table 1 shows the influence of correlation between the objectives on the ranking methods and the different P-ACOs. It can be seen for all algorithms that the size of the non dominated front shrinks (increases) in the homogeneous (heterogeneous) case when the correlation increases. Though this trend can be observed for all algorithms the extent is different. The decrease in the homogeneous case is strongest for CR-P-ACO and STD-P-ACO. The ranking methods seem to help the other algorithms to keep their non dominated front small even when the correlation is low. Opposite to the homogeneous case, the increase in the heterogeneous case is strongest for the algorithms with ranking.

The relative performance of the algorithms can be seen by the fraction their solutions make up of the common non dominated front. Algorithms CR-P-ACO

Table 2. Fraction of solutions from specific algorithm among the top 10% of all solutions concerning a single objective (averaged over the five objectives); homogeneous case (left) heterogeneous case (right)

$|c|$	Pt	WL	W	Rel	CR	STD	Pt	WL	W	Rel	CR	STD
0.2	13.8	16.6	31.7	37.3	0.3	0.3	14.0	16.0	32.5	37.0	0.3	0.2
0.4	15.4	18.0	31.5	34.9	0.1	0.1	15.1	16.9	32.1	34.8	0.6	0.5
0.6	19.8	21.3	26.8	31.7	0.0	0.4	17.2	19.8	26.0	31.3	2.5	3.2
0.8	22.5	23.7	21.6	21.0	0.0	11.2	22.7	23.6	23.7	23.1	3.1	3.7
0.9	14.3	21.0	23.0	17.3	0.0	24.3	23.4	22.6	22.2	24.7	3.0	4.0

and STD-P-ACO are clearly the worst for small and medium correlation $c \leq 0.6$. CR-P-ACO is also the worst for high correlation but in the heterogeneous case its performance improves. STD-P-ACO profits from a high correlation in both cases. In the homogeneous case it is the best algorithm for $c = 0.9$. In contrast, algorithms W-P-ACO and Rel-P-ACO which are the strongest for small and medium correlation $c \leq 0.6$ only have average performance for $c \geq 0.8$. Interestingly, the relative performance of the P-ACOs with Pt and WL are relatively robust against changes of the correlation strength. One possible explanation uses the results mentioned in the Introduction that showed how small and medium correlation increase the size of the Pareto front. Our results imply that the P-ACOs with ranking can handle the larger Pareto fronts better than CR-P-ACO and STD-P-ACO. The results are similar with respect to the best values that were found for one of the objectives (Tab. 2). The main difference is that STD-P-ACO profits from high correlation only for the homogeneous case.

5 Conclusion

The influence of correlation between objectives of multi objective optimization problems (MOPs) on different types of P-ACO algorithms has been analyzed. A simple method to create MOPs with correlated objectives was proposed and applied to the TSP. Theoretically it was shown how Pearson correlations between the objectives can be obtained. It was argued that simple homogeneous cases of correlations between the objectives should be studied for application relevant problems. Experimentally it was shown how different types of correlations between the objectives influence different types of P-ACO algorithms. The results show that ranking methods seem to help algorithms to keep their non dominated front small even when the correlation is low and can make them relatively robust against changes of the correlation strength.

References

1. Xu, Y., Qu, R., Li, R.: A simulated annealing based genetic local search algorithm for multi-objective multicast routing problems. Ann. Oper. Res. 206(1), 527–555 (2013)

2. Jaszkiewicz, A.: Genetic local search for multi-objective combinatorial optimization. European Journal of Operational Research 137(1), 50–71 (2002)

3. Knowles, J.D., Corne, D.: Towards landscape analyses to inform the design of hybrid local search for the multiobjective quadratic assignment problem. HIS 87, 271–279 (2002)

4. Paquete, L., Stützle, T.: A study of stochastic local search algorithms for the biobjective QAP with correlated flow matrices. Eur. J. Oper. Res. 169(3), 943–959 (2006)

5. López-Ibánez, M., Paquete, L., Stützle, T.: Hybrid population-based algorithms for the bi-objective quadratic assignment problem. Journal of Mathematical Modelling and Algorithms 5(1), 111–137 (2006)

6. López-Ibáñez, M., Paquete, L., Stützle, T.: On the design of ACO for the biobjective quadratic assignment problem. In: Dorigo, M., Birattari, M., Blum, C., Gambardella, L.M., Mondada, F., Stützle, T. (eds.) ANTS 2004. LNCS, vol. 3172, pp. 214–225. Springer, Heidelberg (2004)

7. Garrett, D., Dasgupta, D., Vannucci, J., Simien, J.: Applying hybrid multiobjective evolutionary algorithms to the sailor assignment problem. In: Jain, L.C., Palade, V., Srinivasan, D. (eds.) Advances in Evolutionary Computing for System Design. SCI, vol. 66, pp. 269–301. Springer, Heidelberg (2007)

8. Verel, S., Liefooghe, A., Jourdan, L., Dhaenens, C.: Analyzing the effect of objective correlation on the efficient set of MNK-landscapes. In: Coello Coello, C.A. (ed.) LION 5. LNCS, vol. 6683, pp. 116–130. Springer, Heidelberg (2011)

9. Verel, S., Liefooghe, A., Jourdan, L., Dhaenens, C.: Pareto local optima of multiobjective NK-landscapes with correlated objectives. In: Merz, P., Hao, J.K. (eds.) EvoCOP 2011. LNCS, vol. 6622, pp. 226–237. Springer, Heidelberg (2011)

10. Shi, C., Yu, P., Yan, Z., Huang, Y., Wang, B.: Comparison and selection of objective functions in multiobjective community detection. Computational Intelligence (to appear, 2013)

11. Ishibuchi, H., Akedo, N., Ohyanagi, H., Nojima, Y.: Behavior of EMO algorithms on many-objective optimization problems with correlated objectives. In: 2011 IEEE Congress on Evolutionary Computation (CEC), pp. 1465–1472. IEEE (2011)

12. Ishibuchi, H., Akedo, N., Nojima, Y.: A study on the specification of a scalarizing function in MOEA/D for many-objective knapsack problems. In: Nicosia, G., Pardalos, P. (eds.) LION 7. LNCS, vol. 7997, pp. 231–246. Springer, Heidelberg (2013)

13. Ishibuchi, H., Yamane, M., Nojima, Y.: Effects of duplicated objectives in manyobjective optimization problems on the search behavior of hypervolume-based evolutionary algorithms. In: 2013 IEEE Symposium on Computational Intelligence in Multi-Criteria Decision-Making (MCDM), pp. 25–32. IEEE (2013)

14. Brockhoff, D., Saxena, D., Deb, K., Zitzler, E.: On handling a large number of objectives a posteriori and during optimization. Natural Computing Series, pp. 377–403. Springer (2008)

15. Goel, T., Vaidyanathan, R., Haftka, R.T., Shyy, W., Queipo, N.V., Tucker, K.: Response surface approximation of pareto optimal front in multi-objective optimization. Comput. Method. Appl. M. 196(4), 879–893 (2007)

16. Murata, T., Taki, A.: Examination of the performance of objective reduction using correlation-based weighted-sum for many objective knapsack problems. In: 10th International Conference on Hybrid Intelligent Systems (HIS), pp. 175–180. IEEE (2010)

17. Guntsch, M., Middendorf, M.: A population based approach for ACO. In: Cagnoni, S., Gottlieb, J., Hart, E., Middendorf, M., Raidl, G.R. (eds.) EvoWorkshops 2002. LNCS, vol. 2279, pp. 72–81. Springer, Heidelberg (2002)

18. Knowles, J., Corne, D.: Instance generators and test suites for the multiobjective quadratic assignment problem. In: Fonseca, C.M., Fleming, P.J., Zitzler, E., Deb, K., Thiele, L. (eds.) EMO 2003. LNCS, vol. 2632, pp. 295–310. Springer, Heidelberg (2003)

19. Liefooghe, A., Paquete, L., Simões, M., Figueira, J.R.: Connectedness and local search for bicriteria knapsack problems. In: Merz, P., Hao, J.K. (eds.) EvoCOP 2011. LNCS, vol. 6622, pp. 48–59. Springer, Heidelberg (2011)

20. Corne, D., Knowles, J.: Techniques for highly multiobjective optimisation: some nondominated points are better than others. In: Proceedings of the 9th Annual Conference on Genetic and Evolutionary Computation, pp. 773–780. ACM (2007)

21. Ishibuchi, H., Tsukamoto, N., Hitotsuyanagi, Y., Nojima, Y.: Effectiveness of scalability improvement attempts on the performance of NSGA-II for many-objective problems. In: Proceedings of the 10th Annual Conference on Genetic and Evolutionary Computation, pp. 649–656. ACM (2008)

22. Lopez Jaimes, A., Santa Quintero, L., Coello Coello, C.A.: Study of preference relations in many-objective optimization. In: Proceedings of the 11th Annual Conference on Genetic and Evolutionary Computation, pp. 611–618. ACM (2009)

23. Lopez Jaimes, A., Coello Coello, C.A.: Study of preference relations in many-objective optimization. In: Proceedings of the 11th Annual Conference on Genetic and Evolutionary Computation, pp. 611–618. ACM (2009)

24. Moritz, R., Reich, E., Schwarz, M., Bernt, M., Middendorf, M.: Refined ranking relations for multi objective optimization and application to P-ACO. In: Proceeding of the Fifteenth Annual Conference on Genetic and Evolutionary Computation Conference, pp. 65–72. ACM (2013)

25. Drechsler, N., Drechsler, R., Becker, B.: Multi-objective optimisation based on relation favour. In: Zitzler, E., Deb, K., Thiele, L., Coello Coello, C.A., Corne, D. (eds.) EMO 2001. LNCS, vol. 1993, pp. 154–166. Springer, Heidelberg (2001)

26. Garza-Fabre, M., Toscano Pulido, G., Coello Coello, C.A.: Ranking methods for many-objective optimization. In: Aguirre, A.H., Borja, R.M., Garciá, C.A.R. (eds.) MICAI 2009. LNCS, vol. 5845, pp. 633–645. Springer, Heidelberg (2009)

27. Garza Fabre, M., Toscano Pulido, G., Coello Coello, C.A.: Alternative fitness assignment methods for many-objective optimization problems. In: Collet, P., Monmarché, N., Legrand, P., Schoenauer, M., Lutton, E. (eds.) EA 2009. LNCS, vol. 5975, pp. 146–157. Springer, Heidelberg (2010)

28. Angus, D.: Crowding population-based ant colony optimisation for the multi-objective travelling salesman problem. In: IEEE Symposium on Computational Intelligence in Multicriteria Decision Making, pp. 333–340. IEEE (2007)

29. Guntsch, M., Middendorf, M.: Solving multi-criteria optimization problems with population-based ACO. In: Fonseca, C.M., Fleming, P.J., Zitzler, E., Deb, K., Thiele, L. (eds.) EMO 2003. LNCS, vol. 2632, pp. 464–478. Springer, Heidelberg (2003)

30. Deb, K., Agrawal, S., Pratap, A., Meyarivan, T.: A fast elitist non-dominated sorting genetic algorithm for multi-objective optimization: NSGA-II. In: Deb, K., Rudolph, G., Lutton, E., Merelo, J.J., Schoenauer, M., Schwefel, H.-P., Yao, X. (eds.) PPSN 2000. LNCS, vol. 1917, pp. 849–858. Springer, Heidelberg (2000)

Author Index